Archaeology of Mother Earth Sites and Sanctuaries through the Ages

Rethinking symbols and images, art and artefacts from history and prehistory

Edited by

G. Terence Meaden

BAR International Series 2389
2012

Published in 2016 by
BAR Publishing, Oxford

BAR International Series 2389

Archaeology of Mother Earth Sites and Sanctuaries through the Ages

ISBN 978 1 4073 0981 1

BAR Publishing is the trading name of British Archaeological Reports (Oxford) Ltd. British Archaeological Reports was first incorporated in 1974 to publish the BAR Series, International and British. In 1992 Hadrian Books Ltd became part of the BAR group. This volume was originally published by Archaeopress in conjunction with British Archaeological Reports (Oxford) Ltd / Hadrian Books Ltd, the Series principal publisher, in 2012. This present volume is published by BAR Publishing, 2016.

Printed in England

BAR
PUBLISHING

BAR titles are available from:

BAR Publishing
122 Banbury Rd, Oxford, OX2 7BP, UK
EMAIL info@barpublishing.com
PHONE +44 (0)1865 310431
FAX +44 (0)1865 316916
www.barpublishing.com

Table of Contents

Introduction

G. Terence Meaden

Oxford University Department of Continuing Education (Archaeology), Kellogg College, Oxford, UK

It is opportune to visit anew this classical subject—to see how it can be freshly approached in the 21st century by archaeologists and anthropologists using knowledge of the latest discoveries, data and analyses.

Mother Earth is a basic anthropological worldview that reappears in time in various fields on different continents from arguably at least the early agricultural periods onwards. In many societies—as in the Mediterranean and other parts of the world including the contemporary lands of the Amerindians and Hindu Indians—Mother Earth sites and sanctuaries are known or suspected not only from prehistory but from historical times and the modern era too. Additionally, there is material evidence from the nature of art and artefacts, and there can be reputable inferential evidence from the character of associated symbols and images as well. Some sanctuaries are, or can be, identifiable by their shape. At certain sites, as in Guernsey, Brittany, Corsica and other Neolithic lands and Mediterranean islands, rocks or standing stones bear inscribed or sculptured images of anthropomorphic form; whereas sometimes instead megaliths are arranged, or tombs designed, in such a fashion as to imply that a previously-unsupposed culture of Earth Mother belief may be inferred. In other places, as with geometrized man-made objects found at settlements or necropolises, non-iconic patterns often support similar symbolism. Examples are plentiful throughout the many countries bordering the Mediterranean. Even in *northern* Europe, as for instance at an occupation site Villers-Carbonnel on the banks of the Somme, a terracotta fertility-type female figurine, found in a collapsed kiln by Françoise Bostyn in 2011, has been dated to the Early Neolithic 4300-3600 BC.

Although the concept of Mother Earth may seem to favour one gender, specialists in this subject are interested in the wider study of the general anthropomorphization of the world, namely how geomorphology has sometimes supported projections of the human body, male and female.

With these views a session was organised to take place at the Fourteenth Annual Conference of the European Association of Archaeologists in Valetta, Malta, in September 2008.

The papers discussed the various forms of evidence not only from definite sites but from others for which an expression of a divine feminine principle, personified as belief in an Earth Mother or other female deity, may be inferred as possible or sometimes likely—especially where the work is based on new discoveries.

For the present survey additional authors who were extremely interested but unable to attend the session have contributed useful papers as well—above all, archaeologist Professor Dragos Gheorghiu of the National University of Arts, Bucharest, Romania, with a particularly relevant paper. This diversity of authors and heterogeneity of approach yields a broader range of intellectual interpretation and analysis than could be produced by any single researcher.

*

Firstly, there is an influential paper by *Dr.Nikos Chausidis* who explores material examples of the many female images of Mediterranean origin known for the millennia that traditionally were deemed to depict the Mother Earth or Mother Goddess for the Metal Ages BC and early centuries AD and which also reach into Byzantine and Medieval Christian times. For these centuries there can be helpful literary support for the fertility images but the beginnings lay in the more distant, preliterate, less clear Neolithic.

Subsequent papers are ordered into approximate chronological sequence—starting with the Late Upper Palaeolithic and Early Neolithic and passing through all periods into the Christianity of Medieval and later times and concluding with traditional activities in present day Thrace and Hindu India.

In the Upper Palaeolithic sculptured forms of the human female are known widely across Europe from France through to Russia. Their meaning and use are unknown—although much discussed—so suppositions as to implied divinity or fertility charms remain unsupportable on present Palaeolithic/prehistoric evidence. Instead, one may attempt to usefully consider what early peoples of the Neolithic may have thought about landscape. Among the many researchers on this topic one may cite Keith Basso (American Indian tribes), Edwyn C. Krupp for aspects of Earth Mother sites of American Indians) and Christopher Tilley for phenomenologies of European ancient landscapes. Among pioneers in interpreting symbols and images of preliterate prehistoric peoples, Joseph Campbell, Mircea Eliade,

Gordon Fraser and Marija Gimbutas have discussed matters at great length.

In the present report *Dr. Bostjan Odar* of Croatia opens by perceiving aspects of an Upper Palaeolithic landscape in mountainous terrain at the caves of Potocka Zijalka and Divje Babe I. Their landscape localities are considered as well as the context of finds inside the caves. He is led to consider whether the occupiers of the caves had deemed a cosmological principle to be recognisable in the landscape.

Next, for Central Sardinia *Drs. Gradoli and Meaden* summarise aspects of a vast newly-located Neolithic landscape in the hills near Seulo. Pottery style by seriation and human bone analysis by radiocarbon dating led to dates in the fifth and fourth millennia BC. A rare finding is a Neolithic rock painting that seems to depict a masked hunter shooting an arrow at a wild beast. Street dancing by similarly masked men take place even today in a hill town not far away in Central Sardinia. Features of the cliff geology and the water riverine landscape attracted the attention of prehistoric communities, features that by their nature intimate an open interest in the feminine divine. *Dr. Prendergast* considers aspects of the fourth millennium passage grave at Newgrange in Ireland where mounting indirect evidence from artefacts and construction detail help point towards a concept of internal uterine character that could imply hopes of rebirth in a feminine environment. This may intimate feminine divinity to some degree.

Again, writing from Sardinia *Professor Ugas* reviewed by way of illustrated iconography the likelihood of an island ideology, in this case Sardinia, involving what he interprets as a cult of the Mother Goddess. The next papers—by *Terence Meaden*—demonstrate that the third millennium British monuments of Avebury and Stonehenge each has a key standing stone that answers to the grand mythological design strategy that evokes the Marriage of the Gods as known for later classical times. The same concept, moreover, relates to the even older monument of Newgrange. Additionally a literature study with respect to megalithic Avebury reveals that the commonly viewed three-stone cove really comprised five stones. One stone, now missing, was fully recumbent. Another—a fourth standing stone—functioned in front of the Avebury Cove much as at Stonehenge the external Heel Stone visibly casts a summer solstice phallic shadow into the monument—which it still does today as intended 4500 years ago. This study was aided by constructing and setting at the focus of Stonehenge *a full size replica* of the once upright Altar Stone.

Notwithstanding the well-presented paper of Dr. Chausidis on Mediterranean figurines mentioned above, *Dr. Naumov* questioned whether the corporeality of Macedonian female images that he cites were ever considered 'divine'. He wondered whether these Neolithic figurines could have served some alternative, unknown social function in a community context.

Dr. Liveri writes of the concept of the 'Bee Goddess' as being valued as a holy figure from the Neolithic period to the Mycenaean epoch and not that of a cult of a Bee Goddess. In this manner the bee was linked, as also other animal forms, to the worship of a major female deity like a Mother Goddess while expressing expectations for the fertility of animals, crops and other vegetation.

Next, *Dr. Di Fraia* introduces a cult site in Italy where a body-length megalith propped adjacent to a rock face had arguably served as a healing bed in a ritual that in later times Romans called *incubatio*. This is supported by the character and positions of signs and symbols in close proximity that could be fingered while a believer lay on the rock and viewed painted anthropomorphic female figures. The author proposes alternative entities: one representing the feminine principle (Mother Goddess or Mother Earth) with "values of fertility, reproduction and care", and the other, the masculine (Sun) that "combines power, energy, heat, light and life."

Professor Gergova lectured on the evidence (namely, monuments, images, symbols, inscriptions) for the cult of a Mother Goddess by the literate Getae at Dausdava in Northern Thrace in the first millennium BC. Most importantly, *public rituals of that era are preserved and maintained by the local communities even to the present day*. One is a rite involving lying on a megalithic slab that recalls the *incubatio* or hopes of healing discussed by Dr. Di Fraia. Another is the persistence of the tradition of passing garments through the hole of a perforated stone.

Dr. Orhideja Zorova introduced ideas appropriate to the Medieval period and the rise and spread of Christianity. The earlier regard for Mother Earth as a divinity was increasingly reduced to a personification. Because of the superstitious behaviour of ordinary people, the well-loved figure remained, although sidelined by the church to "the margins of official doctrine". Besides archaeological evidence, many clues survive in other media "such as art, religious writings, and architectural and geomorphic forms in some way connected to the concept of Mother Earth." The functions once possessed by Mother Earth were passed to "the Virgin as protector and benefactor of the deceased" and to Sophia as a "Mother of all Creation", i.e. the material world.

Dr. Katya Hrobat used aspects of Slavic folklore, mythology and archaeology with respect to anthropomorphic megaliths, rock cliffs and place names called Baba from Slovenia and Croatia to demonstrate how the divine feminine principle has been preserved throughout all epochs until the present day. Babas relate to lithic images of an elderly wise woman regarding which specific rituals are performed, while Baba traditional stories seem to reach as far back as pre-Slavic times including Greek and Egyptian traditions as well.

Professor Roberta Robin Dods examined a perceived human sensitivity to the sacredness of space and a unifying one-ness of the earth through the 'dust of the ancestors'. In this way she assessed information gathered from the indigenous religions of the Americas as regards ideas of divinity that are centred on the feminine principle linking with what can be termed and recognised as Mother Earth.

The final paper by *Professor Dragos Gheorghiu* is propitious and hugely important. It brings an understanding of, and the use of, the Mother Earth principle right up to date—namely, to modern times and to sacred places where the divine feminine principle continues to be worshipped with deep devotion. In Bangalore in central India geo-anthropomorphs were photographed a few years ago during their ritual creation by a devout field artist through the medium of fresh paint coated on image-bearing rocks. Through immediate acceptance by the waiting, watching village community, the rock site became a *darshana* or natural sanctuary, as Richard Bradley, Dragos Gheorghiu, Mircea Eliade and Joseph Campbell, among scholars, have discussed. This was because the painted stone had then acquired a specific sacredness—hence Earth too—with a quality and reverence bearing the implicit concept of Earth Mother or Mother Goddess. Indeed, as Prof. Gheorghiu writes when explaining the Bangalore reverential painting event as a 'test' experience, the people were at once satisfied that "after the consecration, the rock embodied shakti, the female power and energy." It reminds us, too, that many of the prehistoric megaliths, thousands of years old, that we view in the countryside today—together with rock features of specific types—may formerly have been painted, say with white chalk in Wessex or red ochre as known for Bronze Age Swedish rock art.

*

Several papers in this collection demonstrate how certain, well loved, traditional symbols and practices have a noted tenacity for surviving the ages of time. The concept of Earth Mother—or 'Goddess' or 'Dea Madre', using related definitions encountered in the Mediterranean lands—is associated with some sites and symbols for which we have knowledge through historical scholarship or anthropological interviews with current believers. For others, an Earth Mother attribution—as similarly with ideas of a Universal Ancestor—is for many prehistoric sites only a theoretical concept. This explains why the Earth Mother hypothesis has been put to the test by some of the researchers in this book. Indeed, a new era of rigorous testing has begun. Prof. Gheorghiu closes his timely essay by concluding: "the processes of anthropomorphization, as well as that of epiphany and hierogamy, which were specific for traditional societies, were not totally withdrawn by Modernity, and still could exist unrevealed within contemporary imageries".

Mythical Representations of 'Mother Earth' in Pictorial Media

Nikos Chausidis

University of Skopje, Institute for History of Art and Archaeology, Macedonia.

Abstract. *This paper summarizes our past researches of the pictorial representations of the Mother Earth myth and the separation of the basic iconographical types. Generally, the paper is not geographically cultural or chronologically limited. This means that we approach the phenomenon in its wider aspect, searching for its universal (transhistorical and transcultural) features. This is justified by the simple fact that the Mother Earth phenomenon itself possesses such a character, being universal for the bigger part of mankind. Yet, beside this principal openness, the focus of our research points toward the archaic cultures, i.e. those that had never, or not in a sufficient degree, entered the spheres of the cultures that are today regarded as civilizations. Here we have in mind the cultures of the Neolithic, the Age of Metals and the later centuries BC. We have divided the corpus of the pictorial representations of Mother Earth into several categories based not so much on the appearance but on the basic semiotic concept that generated them: namely (i) A female figure in a determined posture; (ii) Reduction and partition of the female figure; (iii) Geometrisation of the female figure; (iv) Zoomorphisation of the female figure; (v) Woman that appears from the earth.*

Key words*: Archetypes, Birth-Giving Goddess, Mother Earth, pictorial representations.*

Introduction

Within the basis of this research lies the question: *"how could archaic man represent the mythical character Mother Earth in pictorial media"?* The results that are exposed here represent a brief synthesis of our past detailed researches previously published as separate studies in journals or as chapters of monographs.

An answer to the question raised can be offered by the medium of language. If our starting point is the way in which the term and the meaning of the mentioned mythical character is formed inside this medium (lecsem "mother" + lecsem "earth"), rather than within the pictorial media, it too should be constructed in an analogous way—through the merging of the pictorial motifs of both elements: the picture of a mother plus a picture of an earth. Because of this, we believe that in the beginning it would be necessary to determine how archaic man could pictorially represent these two elements.

In pictorial media the function 'mother' could have been represented through a figure of an adult woman with clearly articulated gender features. But this could be accomplished more explicitly and more impressively through the representation of the birth-giving act itself, i.e. the emergence of the baby from the uterus of the woman, or actually the very moment in which she receives the function of a "mother" (Plate I: 6,9). It should not be overlooked that for the members of archaic cultures it was not that simple to represent this scene, because the concepts of the realistic, i.e. naturalistic, articulations of reality were not completely elaborated inside their vicinities. In one such constellation it appeared simpler, as well as more economical, to evoke this complex scene only through the figure of a woman represented stylized/schematized, in some of the birth giving poses—with stretched legs, or a spreading and a bending at the knees (Pl. I: 5,7,8) (Čausidis 2005, 133-137; Chausidis 1994, 155-163). Breastfeeding is the second function that determines a mother. According to this, it could as well encode the function 'mother' in pictorial media in both ways: i.e. either explicitly, as a representation of a woman who nurses a child; or implicitly, as a representation of a woman whose breasts appear emphasized in different ways (through their hypertrophisation, nudity, showing, touching or holding, i.e. through their offering with the hands) (Pl. IX:3,6) (Čausidis 2005, 137, 200-203; Chausidis 1994, 84-89).

How could archaic man pictorially represent the second mentioned element?

It seems that archaic cultures did not develop concepts for objective pictorial representation of the earth as a material, i.e. structure. Nor did they articulate concepts for naturalistic, i.e. realistic, representation of the landscape, and in that way for the earth as its part as well. Unlike us, these cultures did not posses objective, i.e. scientifically based, representations of the shape of the earth, so that every culture formed personal representations regarding this question, based on the characteristics of the local environment and the specific treatment of the culture towards it. According to one of the most typical archaic representations in the universe, the earth was imagined as a square plate above which the circular sky in the shape of an arch or vault arose (Pl. I:1). This form of the planet does not rely on any of its realistic features, but on the inner human perceptive and orientating apparatus, which is constituted along two axes, namely four directions (in front, behind, left, right). Such spatial structures of the universe appear

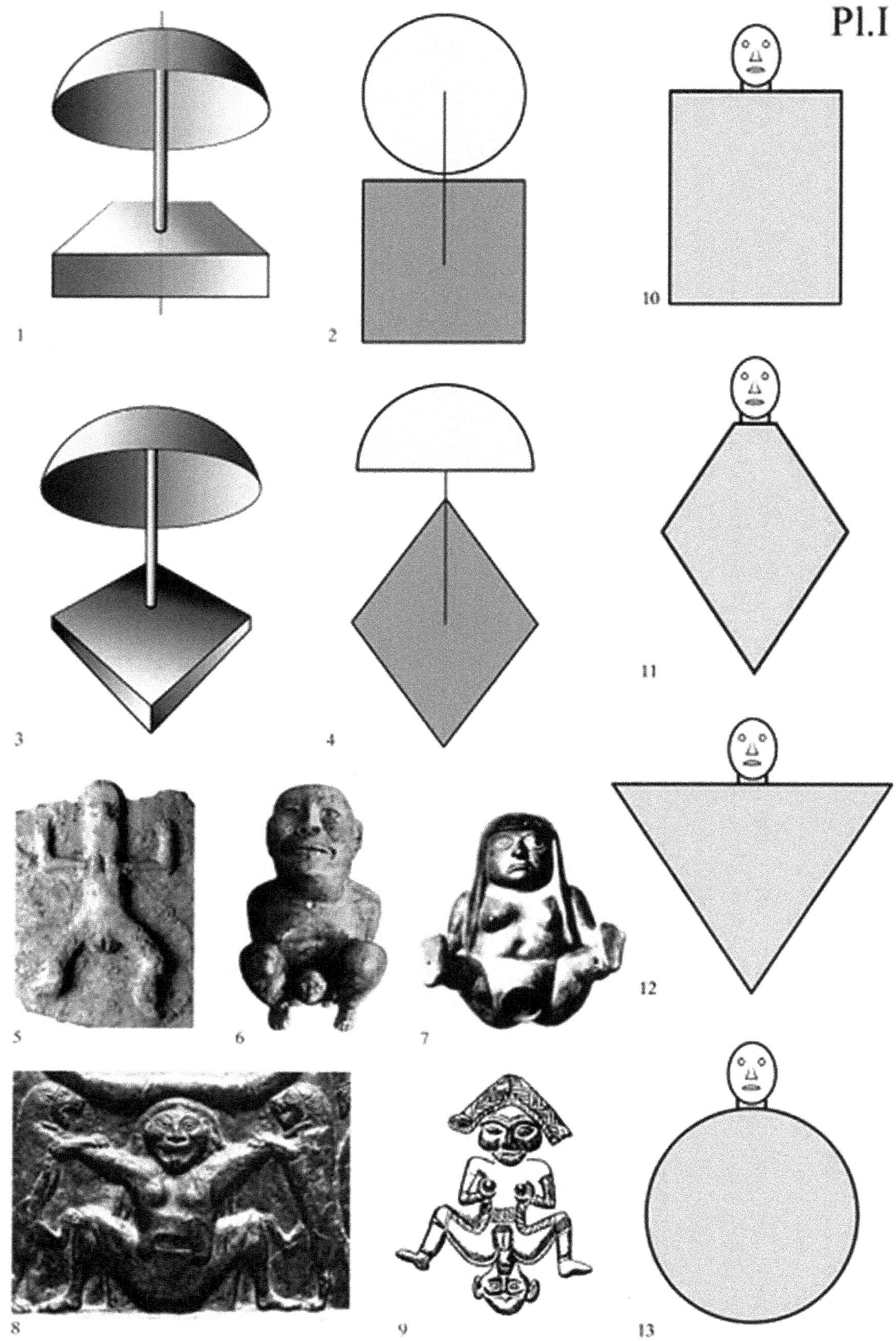

Plate I.
Figs. 1-4. Three-dimensional models of the space and their two-dimensional transpositions (Čausidis 2005, A5, A12); 10-13. Personalization of the geometric pictures/ideograms of the earth (Čausidis 2005, V13); 5. Osijek (Croatia), Neolithic; 6. PreColumbian America; 7. Mochica (Peru); 8. Perugia (Italy), c. 500 BC; 9. Luristan (Iran).

as descriptions in verbal mythical forms, but also in the elements of the material culture such as architecture and sculpture, mostly those having a ritual character. But in the pictorial media these three-dimensional representations had to be compressed into two dimensions—usually through a 'flattening' or 'intertwining' of the two mentioned shapes. As a result, schematized geometrical pictures appeared, in which base the square was shown (= horizontal projection of the earth plate) and above it the circle or the semicircle (= horizontal or vertical projection of the heavenly arch/ vault), both connected through some elongated vertical element (= cosmic axis) (Pl. I:2). There were also variants in which the earth was represented by the rhombus, the reason for which should be looked for in two places (Pl. I:3). On one side, the rhombus, as well as the square is again with four angles, while on the other side, with its shape it resembles an open vulva, thus coding the "female", i.e. the productive, aspect of the earth. The two-dimensional flattened projection of this variant is the rhombus—earth, surmounted by a semicircle—or a circle for the sky, among which a vertical segment stretches, representing the cosmic axis (Pl. I:4) (Čausidis 2005, 30-37).

After we have represented the visual manifestations of the two components that construct the syntagma 'mother-earth', the need appears for yet another summing of the question *"Why is the woman, i.e. the mother, equated to the earth?"* We are convinced that the archaic consciousness (which perceives and valorizes more functionally than morphologically) paid attention to the spatial interference between the two functions of woman and those of the universe. Namely, in both cases, the birth-giving, i.e. the creating (of the child and the plants), is realized by their lower zones, while the nurturing is realized by the upper zones. In this context, the genital organs were equated to the earth from which plants grew, while the breasts from which milk flows were equated with clouds and rain. Because of these functions, there appeared a mythico-symbolical equalization of the earth and the lower part of the female body. This gradually led to their global identification and even to personalization of the earth with a female mythical character with emphasized generative functions. Thus the symbol and the deity Mother Earth was formed, where inside the frames of the oppositional concept the sky was given the husband/spouse character, which, with its dynamical components (sun, rain, light, warmth), fertilizes, i.e. provokes its creativity.

After the elaboration of the basic components that compose the mythical character "Mother Earth", we can approach the elaboration of its basic symbolically-iconographical types.

1. Mother made of earth

The three above-mentioned representations of the mother (as a figure of a woman, of a birth-giving woman, or of a breast-feeding woman) could, within the frames of some archaic culture, represent the goddess Mother Earth, yet not by themselves, but accompanied with the verbal traditions that would denote the given meaning. The presence of the component 'earth' could be accomplished through the articulation of these representations 'using earth', where we literally mean the plastic form made of any kind of earth, like mud or clay, which was afterwards dried or baked, i.e. transformed into ceramic. According to this, we cannot exclude the possibility that Mother Earth was represented by different archaeological objects on which a woman is pictorially represented (in a pose of giving birth or breast feeding, but also in some more usual pose), especially if they are made of unbaked earth/clay or ceramic. If we have in mind that the stone also represents a part of the earth, i.e. belongs to the lower cosmic zone, then this group could potentially enclose the pictorial representations of the mentioned types that were made of stone.

2. Personalization of the geometrical pictures of earth

The square as a picture, symbol and ideogram of the earth encoded only the humans' representations of the form of the earth, i.e. the shape of the "earthly plate" (Pl. I:1,2). The rhombus went deeper, because alluding to the vulva it connoted the generative / creative functions of the earth (Pl. I:3,4) (Čausidis 2005, 93-130). But neither of the two represented one essential aspect—the conceptions that the earth is a living being and that it is a person, a deity. The archaic mythical consciousness spontaneously compensated for this lack in a very simple way—by anthropomorphization of the mentioned geometrical images, i.e. by adding certain elements of the woman's body. One of these ways is realized though articulation of the rhombus and the square with a human (female) head, i.e. face, which will lead to a personalization of these images (Pl. I:10,11). Thus, both geometrical forms receive the character of elements of the body of the represented mythical character, while the added head or face becomes a feature of its personality (Čausidis 2005, 148-160).

Such examples can be identified in different cultures all around the planet. The variants with the square and the rectangle for the time being seem rare and more typical for the prehistoric (Müller-Karpe 1974, Taf. 344; Gimbutas 1989, 238) and the more contemporary archaic cultures (Pl. II:1-4,8,9). They survive in the more advanced cultures from late prehistory and the Ancient Century, but were usually reinvented within the frames of one naturalistic conception. During this, the geometrized (un-anthropomorphic) body of the Goddess-Earth receives the shape of her corpulent torso, which is usually covered with some wide dress, garment, i.e. tunic (Pl. II:5-7,10). We consider that regarding the given detail of the early medieval fibula (Pl. V:3) a compromising solution was reached by using a polygon which alludes to a geometrized earth, but also to a schematized birth-giving pose of the goddess. From the Neolithic to the Antique Period a variant can be followed where the face of the goddess is marked inside the rectangle that represents the earth (Pl. V:2,4,5). The variant with the anthropomorphized rhombus is in a way more convenient. We find her on a sacral stone which was placed in the shrine near Obuhiv

Pl.II

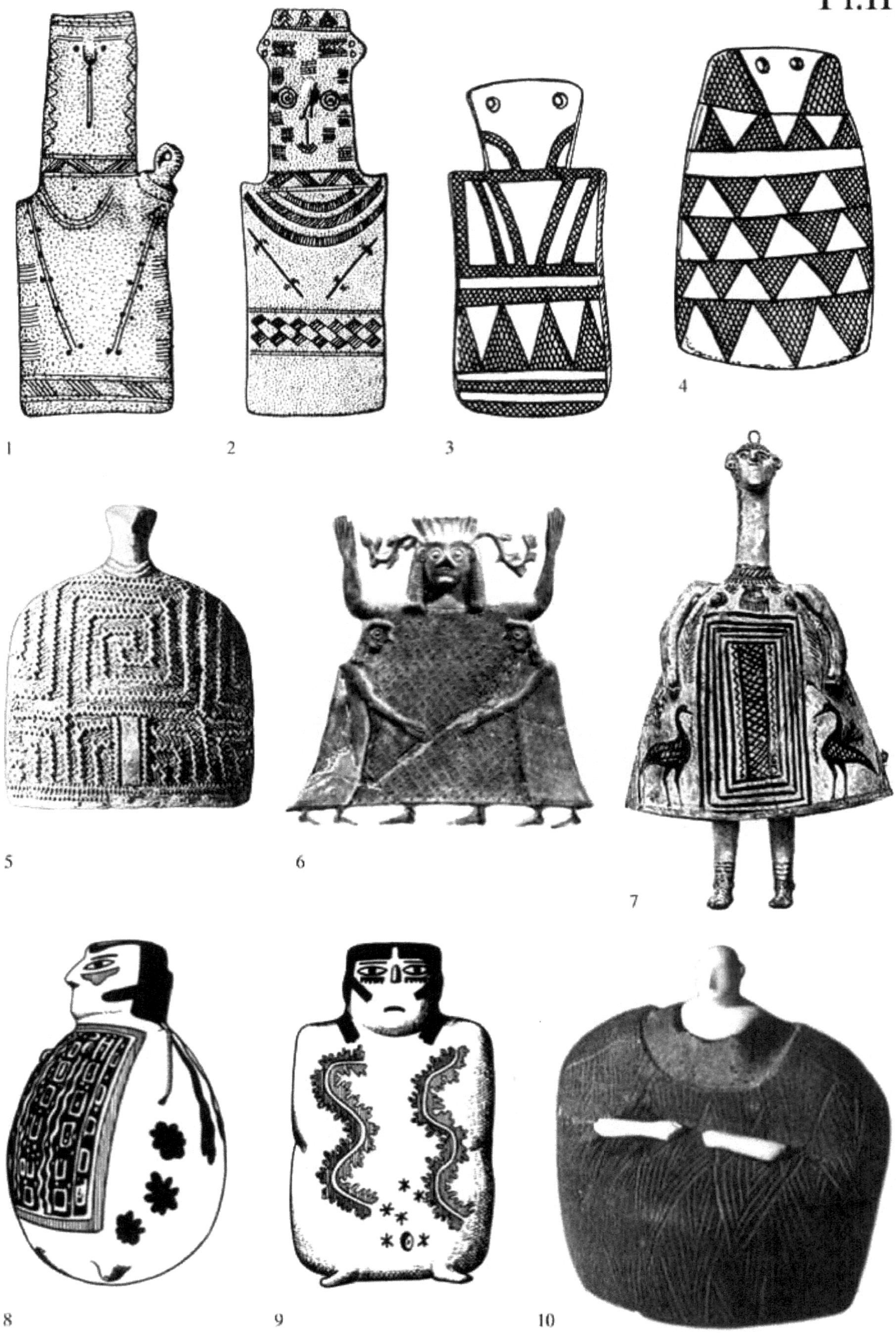

Plate. II
Fig. 1. Lapithos (Cyprus), Eneolithic; 2. Vounous (Cyprus), Eneolithic; 3,4. Cabeco de Arruda (Portugal), Neolithic; 5. Vadastra (Romania), Neolithic; 6. Surroundings of Athens (Greece), 6th-century BC; 7. Boeotia (Greece), Archaic Period; 8,9. Nasca (Peru); 10. Gonur (Turkmenistan), Third millenium BC.

in Ukraine, which existed during the first half of the first millennium AD (Pl. III:1) (Kravchenko 1998). We believe that it represented Mother Earth, where ritual sacrifices presumably took place. Even from ancient written sources we learn about certain stone plates (probably similar to the mentioned one) or about different pieces of stone which symbolized the great Mediterranean Goddesses: Juno, Arthemis, Kybela (Dumezil 1997, 35; Srejović / Cermanović 1987, 203). Rhombic iconographic variants are found in different epochs and regions (examples: PL. III: 2,3). We intercept them on the elongated part of the early medieval two-plated fibulae (6th-7th century) which are connected to German cultures (Kühn 1940; Koch 1988) and Slavic cultures (Korzuhina 1996, PL. 82:1) (Pl. III:4-6). The mentioned significance of these pictures neatly fits into the cosmological iconography of this jewellery, in whose frames its elongated plate represented the lower zones of the universe (Čausidis 2005).

This iconographic conception is also found in two more variants in which the squares alternate with a triangle or a circle (Čausidis 2005, 150, 151) (Pl. I:12,13). We consider that in both cases these geometrical elements also functioned as ideograms for the earth. Similar to the rhombus, the triangle gained this role based on the identification (but this time only functional and not morphological) between the earth and the female genitalia or precisely the Venus area / the pubis as its most visible part. This merging of the triangle and the human (female) head is found on one Neolithic sacrificial altar from Hungary (Pl. IV:8) (Gimbutas 1974, 130; Fig. 108, 109), but also on one category of amulets from the Middle east (Pl. IV:6,7) (Müller-Karpe 1974, Taf. 163; Holmberg 1996, 117). In a more hidden way we also detect it in the structure of concentric triangles which is present on a Neolithic female idol from Bulgaria (Pl. IV:9,10) (Todorova-Vajsov 1993, 207).

It should be pointed that, besides the representations of earth as a rectangle, those in which it was imagined as a circle (e.g. circular plate or some other circular object) are quite common in the world. It should be emphasized that these traditions quite by chance coincide with contemporary knowledge of the circular, i.e. the ball shape of Planet Earth. Reversibly, they are usually treated as most archaic representations in which the concept of orientation by two crossed axes (which will lead to the appearance of the representations of the four-sided earth) has not yet been applied (Čausidis 2005, 16-32). Through numerous variants, this conception is followed on the Eneolithic stone representations from Kultepe (Pl. IV:1-3) (Müller-karpe 1974, Taf. 296; Aydingün 2003, 35-39), where the earthly meaning of the circle is coded by the middle triangle (Pl. IV:2,3) or by other elements which represent "scenes from life on earth" (Pl. IV:1), while personalization was accomplished by one, two or three anthropomorphic heads (= hypostases of the goddess-earth). On the Eneolithic cult vessels from the Cycladic Islands one finds other variants of this same conception (Müller-Karpe 1974, taf. 362; Gimbutas 1989, 102—Fig. 166), where the anthropomorphisation of the earth is not accomplished by adding a head but by metamorphising the circle into a lower part of a female figure which is reduced to its enlarged hips and legs (Pl. IV:4,5). One example from Siros (Pl. IV:4) clearly points that it is the lower part from the body of the Goddess Earth, in the midst of whose hips, inside the wavy sea, floats a boat with paddles. The surrounding zigzag margin—also present on other objects of this type—could represent the mythical river Oceanus which surrounded the lower cosmic zones. The productive function of the Goddess is marked with a vulva (a triangle in the middle), to which in this case is added a pair of branches on the sides as symbols of its fertility (Čausidis 2005, 150, 151).

3. Mother Earth appearing behind the horizon

Cultures possessing a more developed sense of realistic pictorial expression tried to replace the abstract scheme "square / rhombus + female's head" with a more natural variant in which the head or bust of the goddess peeks/ appears/ emerges from the horizontal line which marks the surface of the earth, viz. the horizon (Pl. V:8-10). The appearance of this variant usually understands the loss, i.e. the forgetting of the meaning, of the above-mentioned geometric ideograms of the earth. This conception is confirmed through numerous antique examples in which the name, i.e. identity, of the so articulated goddess can be determined. The Gigantomachy from the Pergam Altar is one of those examples where the figure thus represented can be identified as Gea (Pl. V:9), but there are others which have been identified as Demeter or Tellus (Pl. V:1,2,8) (Baring/ Cashford 1993, 306—Fig.2; Marazov 2003, 243). If by this conception we conceive the form "bust" / "half-long representation", then a hypothesis could be suggested according to which it appeared as a form of representing the Goddess-Earth, whose lower part of the body cannot be seen because she is drowned, i.e. equalized with the earth (Pl.V:1). Such examples can be identified in the frames of Egyptian civilization also (Pl. V:10), but it remains undetermined for us as to whether the emerging head and hands belong to Geb or some other character that personified the earth (Čausidis 2005, 152-154).

4. Mother Earth as an anthropomorphic mountain

The ground surface, when undetermined in form and dimensions, is such that archaic conciseness demands some explanation for it, basing it on how it could project its understandings of the ground as a goddess. Hence, the focus usually falls on the mountain, which erupting from the earth as an object with a material volume and shape, becomes a representation of Mother Earth. The Goddess, as a representation of the mountain and at the same time of the earth, was adopted by numerous cultures. One of the most famous is the Phrygian Kybela. Early iconographical prototypes of these mythical images are not known to us, but we learn about them indirectly through their medieval or modern versions originating from Western Europe (Pl. VI.9;

Pl.III

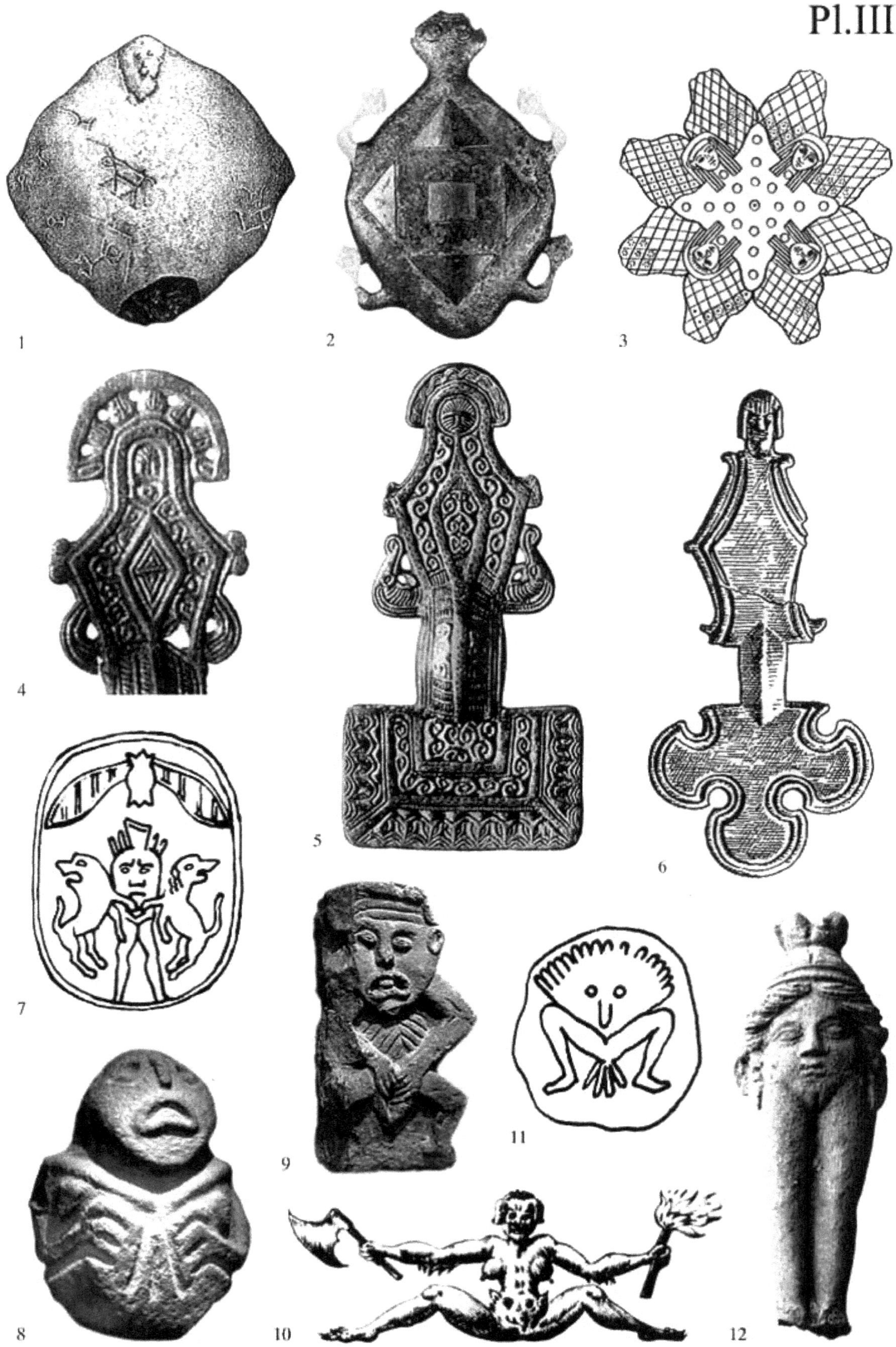

Plate III

Fig. 1. Obuhiv (Ukraine), 1st-5th-century ; 2. Caucasia, 15th-11th-century BC; 3. Phoenicia, 8th-century BC; 4. Dept. Aisne, Early Middle Ages; 5. Müngersdorf, Köln (Germany), Early Middle Ages; 6. Stepancy (Ukraine), Early Middle Ages; 7. Sardinia; 8. Lepenski Vir (Serbia), Neolithic; 9. Ireland, Early Middle Ages; 10. Illustration from the manuscript Openheim, 17th-century ; 11. Zakro (Crete), Mycenian period; 12. Priene (Asia Minor), 5th-century BC.

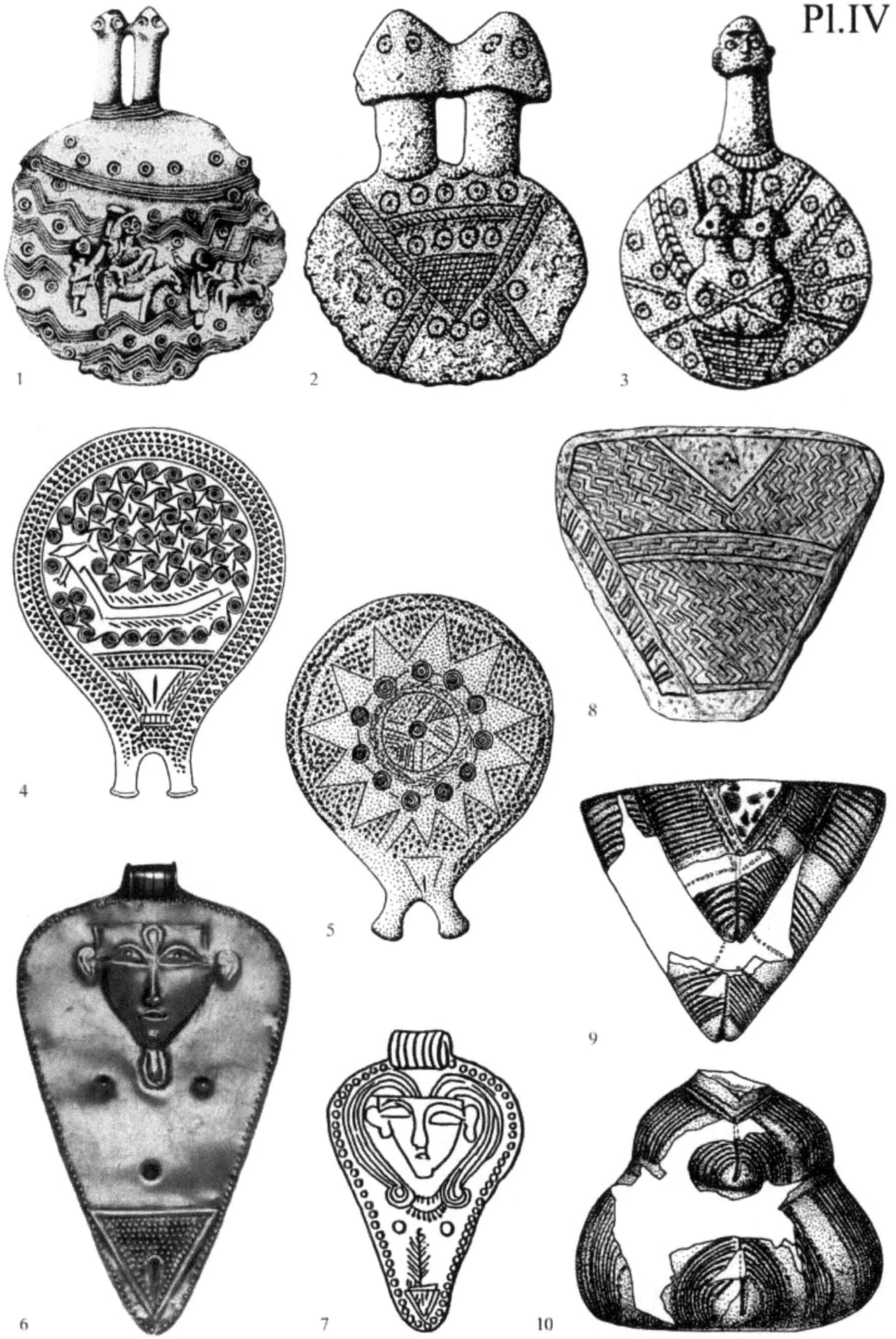

Plate IV

Figs. 1,2,3. Kültepe (Turkey), Eneolithic; 4,5. Syros (Greece), Eneolithic; 6. Tel el Adjuna (Israel); 7. Ugarit, 1400 BC; 8. Kökenydomb (Hungary), Neolithic; 9,10. Černa (Bulgaria), Neolithic.

Pl.V

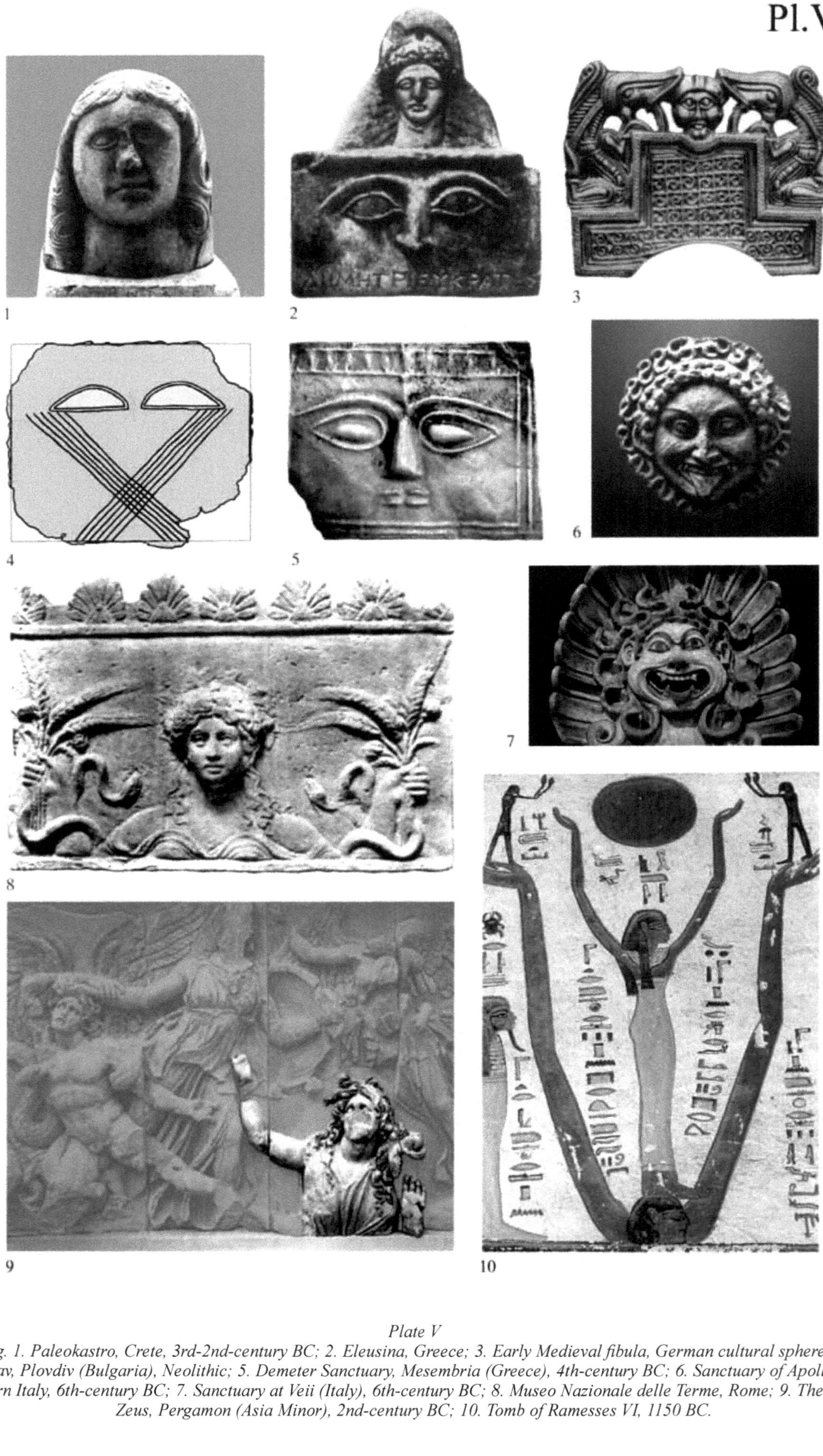

Plate V

Fig. 1. Paleokastro, Crete, 3rd-2nd-century BC; 2. Eleusina, Greece; 3. Early Medieval fibula, German cultural sphere; 4. Dolnoslav, Plovdiv (Bulgaria), Neolithic; 5. Demeter Sanctuary, Mesembria (Greece), 4th-century BC; 6. Sanctuary of Apollo Alaios, Southern Italy, 6th-century BC; 7. Sanctuary at Veii (Italy), 6th-century BC; 8. Museo Nazionale delle Terme, Rome; 9. The altar of Zeus, Pergamon (Asia Minor), 2nd-century BC; 10. Tomb of Ramesses VI, 1150 BC.

10). Although in most examples this character displays features of the Virgin, there are other examples where it still goes under the name of Tellus – the ancient Goddess-earth (Lowden 2003, 22-29). In the examples indicated she is metamorphosed into a mountain, where in some cases the nurturing functions typical for the Mother Earth are accentuated (Čausidis 2008, 275-284; Zorova 2009).

5. Phytomorphization of the female figure

Plants as inhabitants of the earth are often used in pictorial media to represent and symbolize the lower universal zones. Related (whether merged or metamorphosed) with the lower parts of the figure of the birth- giving goddess, they suggest an equality with the earth and thus with the fertility of this character. In a large part of the ancient world the picture of a goddess whose spread legs are transformed into different blossomed plants with branches, leafs, flowers and fruits, is very popular (Holmberg 1996; Marazov 1992, 228-234). We offer several ancient examples from Scythia, Thrace and Macedonia (Pl. VI:5-8), for which we believe they represented Mother Earth. The same meaning could also be given to the representations where the whole figure of the woman is phytomorphized, i.e. transformed into a tree or some other plant, where it often keeps the global contours of a woman represented in a birth giving pose (Pl.VI:1-4) (Chausidis 1994, 167-176; Čausidis 2005, 143, 144, 161-163).

6. Zoomorphization of the legs of the birth-giving goddess

As we have seen, for archaic man it was very difficult to represent and determine space with its three spatial levels—earth, ground and sky—in a pictorial level. The challenge was ably solved by the way that every zone was complemented with animals as their most typical inhabitants. Thus snakes and other reptiles came to mark the underground, while mammals represented the ground and birds marked the heavenly zone. In the same way the figure of the Great Goddess which comprises the whole universe with her body was spatially encoded, the most suitable animals for every zone placed on or by her figure (Pl.VII:1) (Čausidis 2005, 144-146). In this case we are extremely interested in the lower parts of the goddess' body (the birth-giving organs, the abdomen, the hips and legs) which are equated with the lower spatial zones (earth, underground and earthly waters). Within the frames of this concept her legs, spread in a birth-giving pose, are metamorphosed into snakes because of the presence of these animals as symbols of earth and underground. In some cases they turn into snakes' tails (Pl. IX:5), while in others into heads, namely snakes' protomes (Pl. VII:7-9). The fish, as an inhabitant of the lower universal zones (the earthly waters), also participates in this concept of zoo-metamorphization of the birth-giving Goddess—Earth. The result becomes the iconographical type of goddess with spread-open legs which end in a form of fishes flippers; this is clearly confirmed as a specific mythical character in numerous world mythologies (Pl. IX:1,2,4). There are variants where both legs merge into a unique back part with a flipper (Pl. IX:4), but examples also exist where the fishes' pair are not entirely assimilated into the goddess' body (Pl. IX:1). In different cultures the legs of the goddess are metamorphosed into protomes of different animals, which inside the local surroundings, possessed the functions of chthonic symbols or were given some more general sacral character (a horse, a goat, birds of prey etc.) (Pl. VII; PL.VIII) (Čausidis 2005, 170-191; Chausidis 1994, 184-190).

The first of the mentioned variants – the one with snakes' legs is frequently found within the Scythian cultural sphere. The ideal example of this type of deity is the Scythian "Goddess with Snakes Legs", which is determined by historical sources as clearly characteristic of a chthonic deity bearing features of Mother Earth. Despite this description, in archaeological material (like luxurious appliqués made of gold and precious metals), her twisted legs usually do not end with snakes' protomes, but with those of other animals (goats, rams, birds and undetermined species) (Pl.VII:2,3). Variants without protomes are not rare, as well as those with two or more pairs of zoo-metamorphized legs (Pl. VII:2-6,10). Several elements point to the chthonic character of these protomes , viz. legs. *On most examples a disc supplemented with rays appears between them (Pl. VII:6,10) which we believe marks the sunrise, pronouncing its birth from the womb of the Mother Earth.* On the appliqué from Vettersfeld (Pl. VII:12), the pair of animal's protomes emerge from the back flipper of one macro-cosmic fish, on whose giant body other animals are expressed as representatives of the three cosmic zones (fishes, land animals, birds). We suggest that the hawk placed between these protomes may be understood as a symbol of the Heavenly God who is represented at the moment of fertilization of the Great Mother. *This is the hieros gamos* (Čausidis 2005, 174-176).

Examples of this iconographical type also appear in the Thracian area and Macedonia (Pl. VII:4,7,9), where the zoomorphic component of the birth-giving goddess is not very clearly determined. The variant with snakes` protomes is especially present on the antique bronze craters from Macedonia, Italy (Pl. VII:7-9), as well as on the famous example from Vix (France). There are clear indications that the Greek manufacturers of these vessels shaped this goddess according to the mythical conceptions of the local buyers, for which these vessels were intended. According to some sources, they probably represented the goddess Mother Earth, the founder and protector of the local ethnic groups. We have placed a hypothesis that the paradigm of the Macedonian examples could be Harmony—the wife of the mythical Cadmo (Čausidis 2005,176-179).

Such a zoomorphised goddess is also present on the famous bronzes from Luristan – precisely on the cultic objects that experts call "standards". In the intertwining of several parallel iconographic layers, deposited by the centurial existence of these objects, one which suits to the

Pl.VI

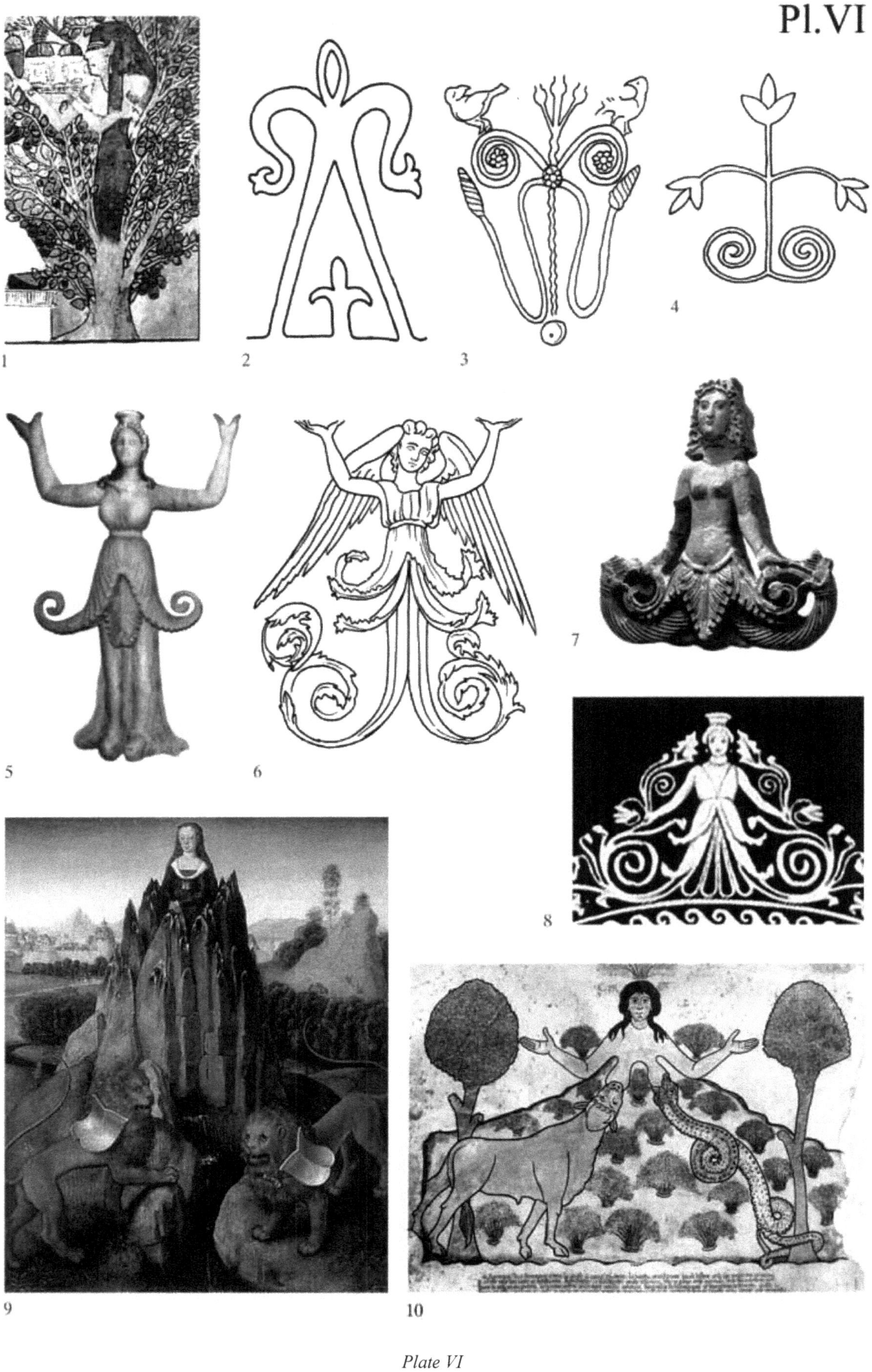

Plate VI

Fig. 1. Thebes (Egypt), 1290 BC; 2. Imotski, Dalmatia, Croatia, Middle/New age; 3. Mainz (Germany), Roman period; 4. Folk embroidery, Slavic cultural area; 5. Tomb in Sveshchari (Bulgaria), 3rd-century BC; 6. Etrurian civilization; 7. Panticapes, Krym, (Ukraine), 1st-century; 8. Pella, (Greece), Helenistic period; 9. Mountain Mother, Allegory of Chastity, by Hans Memling, 1475 AD; 10. Exultet roll, Rome, ca. 1300 AD.

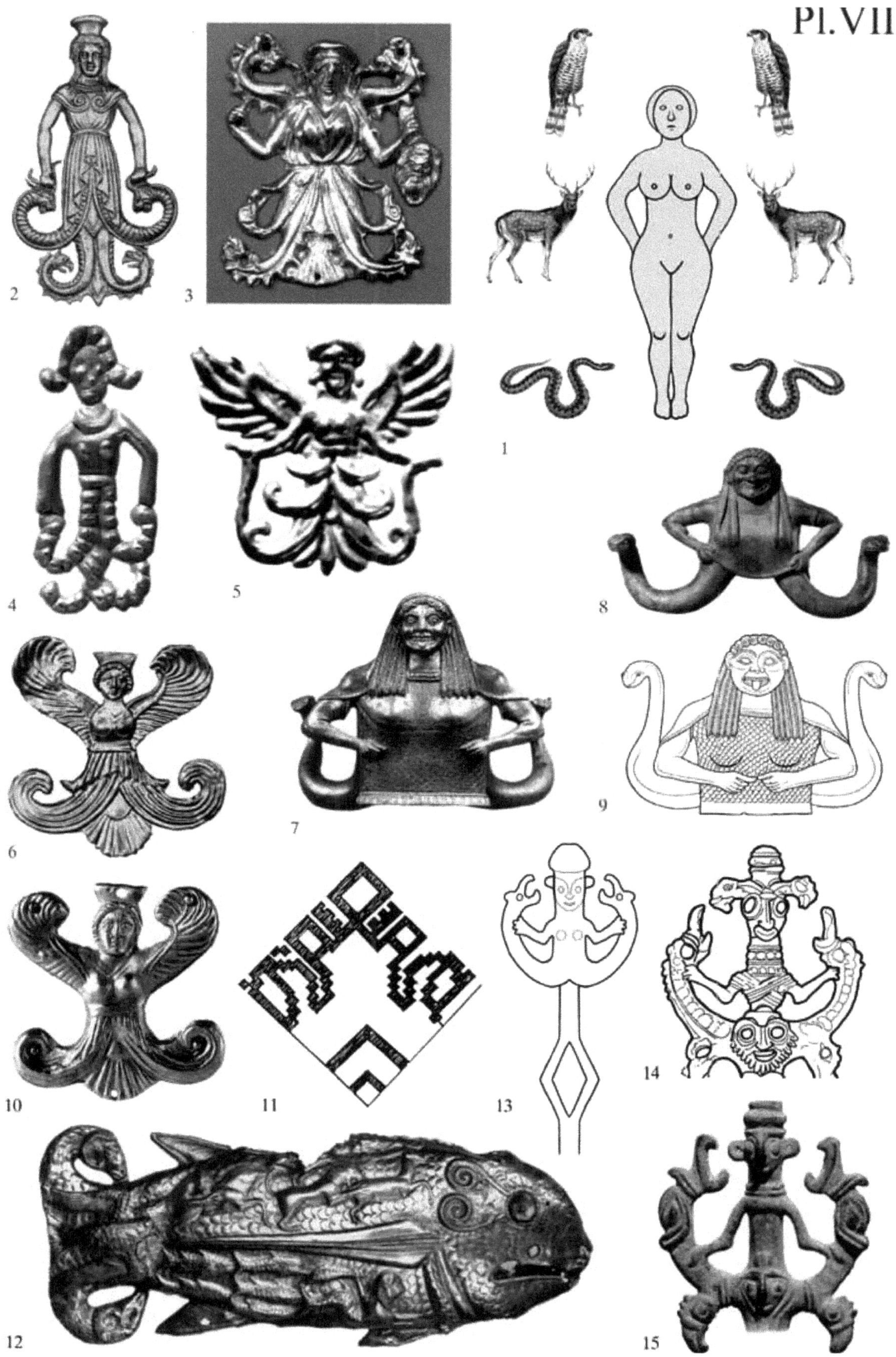

Plate VII

Fig. 1. Cosmisation of the female figure through animals. Metal application, Scythian culture: 2. Bol'šaja Cimbalka; 3. Kul-ob; 5. Hersoness; 6. Aleksandropolj; 10. Hersoness; 12. Vettersfeld. Applicatons for bronze vessel (6th-5th. cent BC.): 7,9. Trebenište (R. of Macedonia); 8. Armento (Italy); 11. Folk embroidery, 20th-century (R. of Macedonia); 13, 14, 15. Luristan Bronze (Iran), first half of the first millenium BC.

here-noted iconographical type (a goddess who holds her spread legs that end like animals` protomes) can be detected (Pl. VII:13-15). Her birth-giving function is also coded with the motif of an open vulva which is represented in the lower zone of the objects (Pl. VII:13) (Čausidis 2005, 60-67, 176).

The figure of the goddess with zoomorphised legs is also clearly followed on the early medieval jewelry from Eastern, Central and Southeastern Europe. She is especially clearly present on the Slavic fibulae of the Dnepr type (Pl. VIII:5-10). On the elongated plate, which represents the lower universal zones, in different prototypes appear specific iconographical variants of this mythical character. Most represented are the horse protoms, goat protoms and those of some other unidentified animals, where in some cases metamorphoses of the horse protomes into birds` protoms can be followed. The female gender of the figure is also marked with the two perforations which signify her breasts (Pl. VIII:5,7,9). On this and on some other medieval examples of jewelry we can also identify a variant in which the hands / arms of the figure are zoomorphised in the same way (Pl. VIII:16). The iconographic similarity of these and the presented Scythian examples is very indicative (compare (PL. VIII:5-10 with PL. VII:2,3,5,6,10), especially if we have in mind that the partial core of distribution of these fibulae matches with the ancestry Scythian territories (Čausidis 2005, 195-197; for this jewelry: Rodnikova 2006; Rodnikova 2006a). The same motif, this time with legs in a form of birds` protoms, also appears on one other similar type of early Slavic fibulae. That is the type named "Maros Gambas Pergamon", which is present from Romania and Hungary, through Macedonia, all the way to Asia Minor (Pl. VIII:1-4). Within the frames of this type appears a variant with one (Pl. VIII:3,4) and with two pairs of birds` protoms (Pl. VIII:1,2), which also corresponds well to the represented Scythian and Thracian examples (Pl. VII:2-5) (Čausidis 2005, 193-195; for this jewelry: Gavrituhin 1991, 131-134).

The mythical character of the birth-giving woman with zoomorphised legs is very present between the Eastern Slavs even after their Christianization. It appears on a category of banded bracelets (Pl. VIII:13) where its representation points to close relations with the mentioned early medieval fibulae, but also with the much older Scythian examples (compare with Pl. VII:3,5). The birth-giving woman with snakes` legs, completed with different animal heads, is present on the reverse side of one category of Russian medieval amulets (known as "zmeeviki") (Pl. VIII:14,15), usually combined with figures of Christian saints, represented on the obverse. She is often accompanied with a written magical formula in which the personalized uterus (Histera) is mentioned as an element with certain negative connotations (malady, death). These examples also show evident relations to the older antique examples (Čausidis 2005, 180-181; Rybakov 1987, 652-656, 691-736; Nikolaeva Černecov 1991).

Although sporadically the figure with zoomorphised legs also appears within the sphere of the medieval German cultures, here it often receives masculine features (Pl. VIII:11). As a motif she is also represented in Romanic church plastic (Pl.VIII:12). Very clear traces of the here-represented iconographical types are followed even in the folklore ornaments of some European and Asian populations, for example those from Siberia, Caucasus and the Slavic regions (Pl. VII:11; Pl. IX:10-12) (Čausidis 2005, 180-184)

7. The mouth of death

What is symbolized by the frightening mouth of Gorgon-Medusa (Pl. V:6,7)—the character often represented like some of the above-mentioned iconographical types (birth-giving pose, spread-open snake legs) (Pl. I:8; PL. VII:7-9)? We are also familiar with earlier and later examples where the woman in a birth-giving pose is represented with a grotesque open mouth, accented teeth and thrusting tongue (Pl. I:6,8; Pl. III:8,9). We follow her from a beginning in Lepenski Vir (Pl. III:8), through the Hindustan Kali all the way to the western European Sheila-na-gig (Pl. III:9). There are indications that the mouth represented in such way associates to yet another opening, located in the opposite end of the woman's body—the vulva. We also find it in the previously mentioned Luristan bronzes, where on the genital zone of the birth-giving woman, between her zoomorphized legs there is a representation of yet another head –grotesque and with wide open eyes (Pl. VII:14,15). The explicit manifestations of this allusion is represented by the mythical character Baubo, whose head and mouth are transported in the area of her stomach and genitalia (Pl. III:7,11,12). It is thought that this character, present in the Mediterranean and in Europe from the Mycenaean period to the New Century, represented a personalization of the vulva. Inside this change of the female's openings the negative aspect of the vulva is coded and with this the negative aspect of the Birth-giving Goddess-Earth. This is "vagina dentate"—the aspect of death, constituted as the swallowing, eating and destroying of life. The word is about a function diametrically contrasted to that of the vulva. It symbolizes the destructive aspects of Mother Earth which always takes back that which she once created as a condition for its perpetual animation (Marler 2002; Neuman 1974; Devereux 1990; Chausidis 2005, 138, 139, 190, 226-233, 281, 377).

8. Frog—birth-giving woman—earth

Apart from the snake, the frog is yet another representative of the reptiles, which very often represents or accompanies the figure of the Goddess-Earth. On Siberian objects (calendars, shamanistic drums) which represent the universe, the frog is placed in a central position and stands for the earth, while the circle around it—the cosmic waters and / or the sky—with the phases of the temporal cycle (Pl. IX:14,15). The reason is the fact that it is both an inhabitant and representative of the lower universal zones. The very shape of the body of this animal, with its muscle and

Pl.VIII

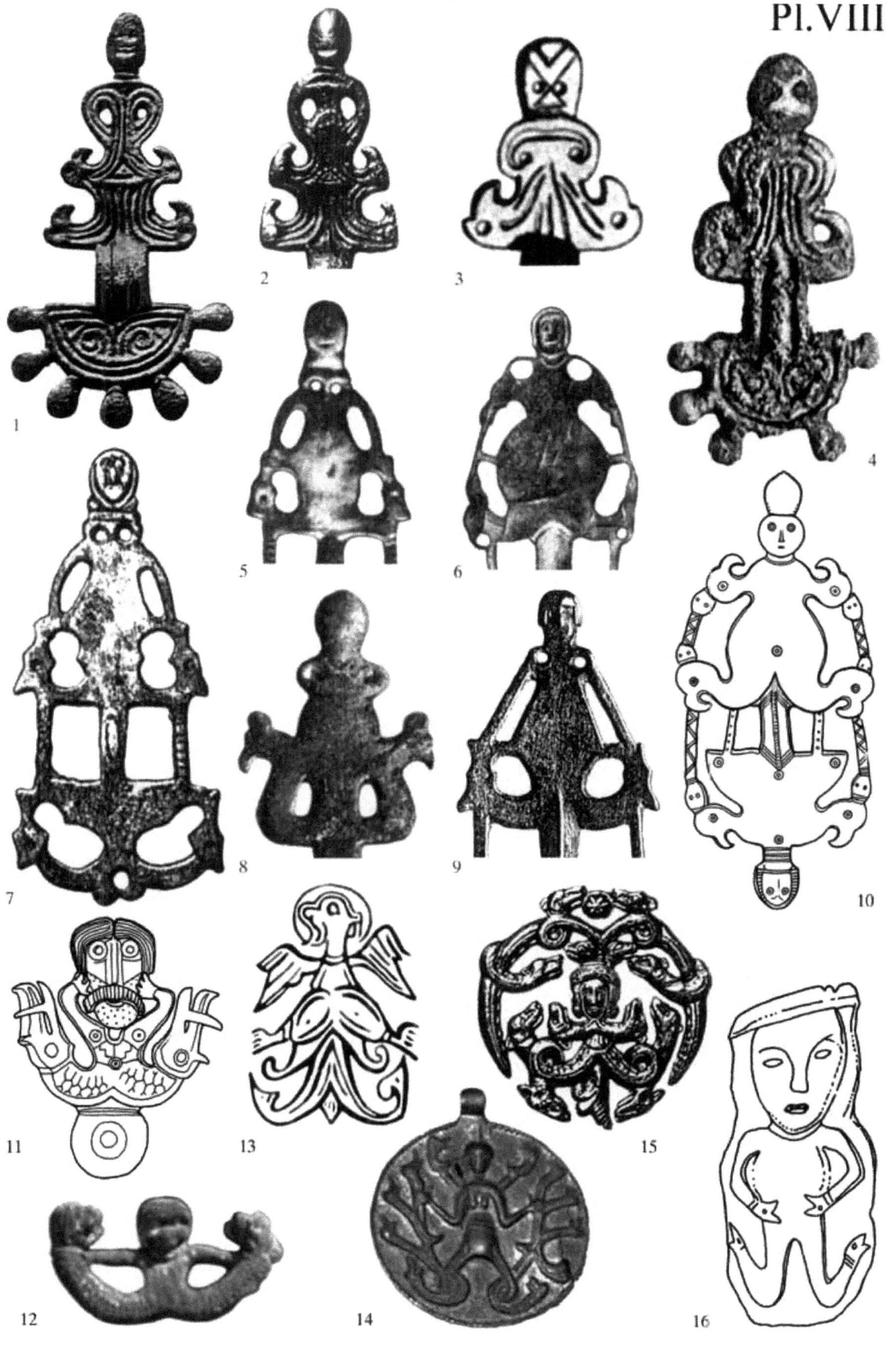

Plate VIII

Early Medieval fibulae: Fig. 1. Maros Gambas (Romania); 2. Tiszafüred (Hungary); 3. Daumen; 4. Pergamon (Asia Minor); 5,6,8,9. Čigirin (Ukraine); 7. Borovka; 10. Blažki (reconstruction of the iconography – Čausidis 2005, D31:3). 11. Aker (Norway), Middle Ages; 12. Church of St. Mary, Poatje (France); 13. Bracelet, 12th-century, National Historical Museum, Kiev (Ukraine); 14. Amulet, Medieval Russia; 15. Amulet, Belgorod, Middle Ages; 16. Application, Kip, Floath of the River Irtyš, Siberia.

Pl.IX

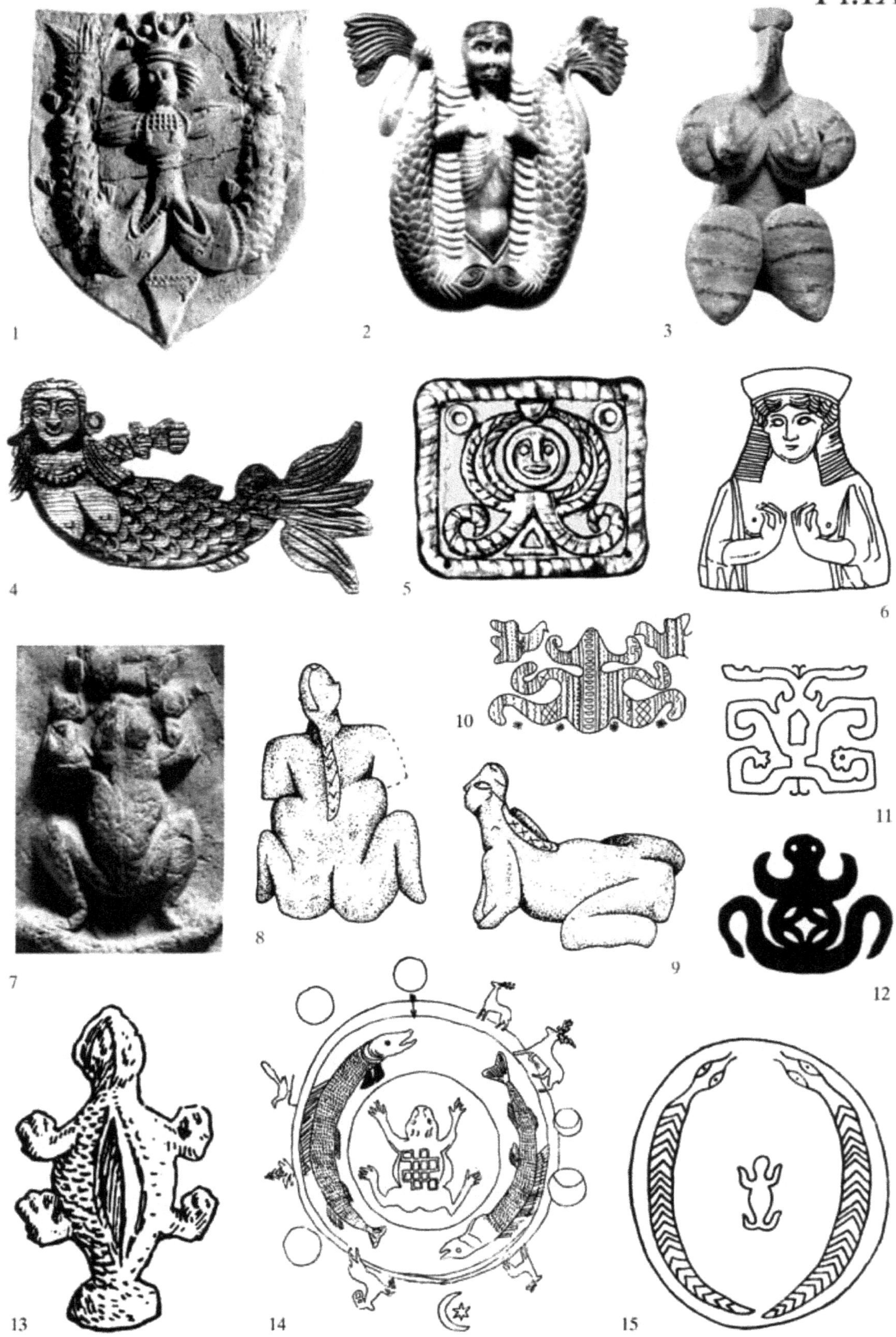

Plate IX

Fig. 1. Prague (Czech Republic), 15th-century; 2. San Pedro di Galigans, Herona; 3. North Mesopotamia, 5th-millenium BC; 4. Floath of the River Volga, Ethnography; 5. Komarovo, Sarmatian Culture; 6. Boeotia (Greece) 5th-century BC; 7. Mesopotamia, 2nd-millenium BC; 8,9. Hacilar (Asia Minor), Neolithic; 10,11. Folk embroidery, Russia; 12. Ornament, Ethnography of the ethnic group Nivhi, Siberia; 13. Votive figurine (Germany), Ethnography; 14. Calendar, Ethnography, Siberia; 15. Leather ornaments, Ethnography of the ethnic group Udegejcy, Siberia

spread legs, alludes to the pose of the woman during birth-giving and coitus (Pl. IX:7, compare to 8,9). It is through this similarity that we should look for one of the basic reasons because of which in many cultures there appears a direct equalization between frog and vulva, between frog and woman, and between frog and birth-giving goddess (Čausidis 2005, 165-167, 209-211; Chausidis 1994, 176-181). Because of these reasons throughout the entire world the frog is present on jewellery, amulets and other ritual objects, which are often connected to the spheres of fertility and birth-giving (Gimbutas 1989, 251-256). In the Balkans and Middle Europe the following ritual still exists: women without children use wax to produce votive statuettes in the form of a frog (Pl. IX:13) and they leave them at certain ritual places, believing that in this way they will conceive and will have a successful delivery (Kus-Nikolaev 1928).

References

AYDINGÜN, S. (2003). *Mysterious Women of the Bronze Age*, Istanbul: Yapi Kredi.

BARING, A., CASHFORD, J. (1993). *The myth of the Goddess*, Penguin Books.

CHAUSIDIS, N. (1994). *Mitskite sliki na Južnite Sloveni*, Skopje: Misla.

ČAUSIDIS, N. (2005). *Cosmological Images: Symbolisation and Mythologisation of the Cosmos in the Pictorial Medium.* Vol. 1 text, Vol. 2 - illustrations and catalogue).

ČAUSIDIS, N. (2008). Mythologization of the Mountain (A Diachronic Survey of Examples from Macedonia and the wider Balkan region), *Memory & Art. Interpretations. Vol. 2, ed. by K. Kulavkova*, Skopje: MANU, 2008, 261-303.

DEVEREUX, G. (1990). *Baubo - La vulve mythique*, Paris (*Bauba - mitska vulva*, Zagreb: August Cesarec.

DIMEZIL, Ž. (1997). *Drevna rimska religija*, Sremski Karlovci - Novi Sad. (Dumezil G. *La religion romaine archaique*, Paris: Payot).

GAVRITUHIN, I. (1991). Paljčatye fibuly pražskih pamjatnikov Podnestrovja, *Drevnosti Severnogo Kavkaza i Pričernomorja*, Moskva, 127-145.

GIMBUTAS, M. (1974). *The Gods and Goddesses of Old Europe*, London: Thames and Hudson.

GIMBUTAS, M. (1989). *The Language of the Goddess*, New York: Thames and Hudson.

HOLMBERG, U. (1996). *Der Baum des Lebens (Göttinnen und Baumkult)*, Bern: Edition Amalia, 1996.

KOCH, A. (1998). *Bügelfibeln der Merowingerzeit im westlichen Frankenreich (T.1,2)*, Mainz, 1998.

KORZUHINA, G. F. (1996). Klady i slučajnye nahodki veschej kruga „drevnostej antov" v srednem Podneprovje (katalog pamjatnikov), *Materialy po arheologii, istorii i etnografii Tavrii*, vyp. V, Simferopolj: Tavrija, 352-705.

KRAVčENKO, N. (1998). Arheologični pamjatki slovjanskogo jazičnitctva (za materialami doslidžennj u Kijvskomu Podniprovji), *Medobori i duhovna kuljtura davnih serednjovičnih Slovjan*, Ljviv, 78-88.

KÜHN, H. (1940). *Die germanischen Bügelfibeln der Völkerwanderungszeit in der Pheinprovinz*, Bonn, 1940.

KUS-NIKOLAJEV, M. (1928). Votivi nerotkinja, *Etnolog*, II, Ljubljana, 36-45.

LOWDEN, J. (2003). Illuminated Books and the Liturgy: Some Observations, *Objects, Images, and the World: Art in the Service of the Liturgy (ed. C. Hourihane)*, Princeton: Princeton University / Princeton University Press, 2003, 17-53.

MARAZOV, I. (1992). *Mit ritual i izkustvo u Trakite*, Sofija, UI Sv. Kliment Ohridski.

MARAZOV, I. (2003). *Hudožestveni modeli na drevnostta*, Sofija: Nov blgarski universitet, 2003.

MARLER, J. (2002). An archaeomythological investigation of the Gorgon, *Styudia in honorem Prof. Ivani Marazov*, Sofija: Anubis, 59-75.

MÜLLER-KARPE, H. (1974). *Handbuch der Vorgeschichte*, Band III (Kupferzeit), München: C. H. Beck'sche Verlagsbuchhandlung.

NEUMANN, E. (1963). *The Great Mother—an analysis of the archetype*, Princeton University Press.

NIKOLAEVA, T. V., ČERNECOV, A. V. (1991). *Drevnerusskie amulety - zmeeviki*, Moskva: Nauka.

RODNIKOVA, V. E. (2006). *K voprosu o tipologičeskom razvitii antropozoomorfnyh fibul (prostye formy)*, Rossijskaja arheologija, 2006/3, Moskva, 41-51.

RODNIKOVA, V. E. (2006a). *K voprosu o tipologičeskom razvitii antropozoomorfnyh fibul (složnye i dvuhplastinchatye formy)*, Rossijskaja arheologija, 2006/4, Moskva, 50-63.

RYBAKOV, B. A. (1987). *Yazyčestvo drevbej Rusi.* Moskva, 1987.

SREJOVI, D., CERMANOVIĆ, A. (1987,). *Rečnik grčke i rimske mitologije*, Beograd: Kultura.

TODOROVA, H., VAJSOV, I. (1993). *Novo-kamennata epoha v Blgarija (krajat na sedmo – šesto hiljadaletie predi novata era)*, Sofija, Nauka i izkustvo.

ZOROVA, O. (2012). Medieval Simulacra of the Mother–Earth in the Christian Traditions, *An Archaeology of Mother Earth Sites and Sanctuaries: Art and Artefacts, Symbols and Images From History and Prehistory.* Published as pages 325 to 332: http://www.ni.rs/byzantium/doc/zbornik8/PDF-VIII/23%20Orhideja%20%20Zorova.pdf accessed 20 April 2012

Perceptions Of The Landscape: At the Palaeolithic Sites Potočka zijalka and Divje babe I, Slovenia

Boštjan Odar

ARC, Palaeolithic Research Centre, b_odar@yahoo.com
Regentova 2, 2000 Maribor, Slovenia

***Abstract**: It is proposed that Potočka Zijalka and Divje Babe I were two centres of religious thought for people who visited them. The position of these caves in the landscape and the context of the finds inside the caves denote a common attitude of a special animistic relationship of humans to cave bears. At a higher level of interpretation, it is possible to detect a geo-solar cosmological principle in both cases. A geo-solar or Earth-Sun principle is incorporated into cosmologies all over the world and is even deeply buried in Christianity.*

***Key words**: Animism, bears, Divje Babe I, Earth-Sun princlple, Potočka Zijalka,*

Introduction

Archaeologists incessantly encounter the past at their work. Questions about the commencement of some phenomena usually remain unanswered, although it is our aim to explain past events with methods appropriate to archaeology. Human beings are, as with all other forms of living and nonliving nature on our planet, unique in their entity. The deeper in time that we descend into a labyrinth of the human past the more the phenomena are hidden. Sometimes we have an opportunity to extract some seemingly insignificant point from the past that waits for a long time before being joined by the next one. By stringing minor advances we try to build up and interpret the story about the past, as it seems to have elapsed.

This research is concerned with Middle and Upper Palaeolithic perceptions of the landscape as indicated by evidence from two atypical Slovenian Palaeolithic cave sites, Potočka zijalka and Divje babe I (Fig. 1). This is based on the unusual find contexts in these caves which both occupy outstanding positions in the landscape. It leads to an interpretation of the sites as centres of the cosmological worldview of the Palaeolithic people in this region. Here it is also suggested that this cosmological view evolved further in much younger periods of human history.

The finds at potočka zijalka and symbolic interpretation

On the western side of Olševa mountain the cave Potočka Zijalka is positioned at an altitude 1630 m (Gauss Kruger co-ordinates x = 5145080, y = 5476960). The cave was named after a previous owner of the cave, Potočnik. Today the owner is Pokrajinski Muzej Celje (Museum of Celje) (Brodar and Brodar 1983; Pacher *et al.*, 2004).

The cave Potočka Zijalka is well known among people and mountaineers from both sides of the state border between Slovenia and Austria (Fig. 1). J. C. Grosz from Bela (Bad Vellach, Austria), who was investigating caves, in autumn 1926 visited Potočka Zijalka several times and began illegal excavations there. He dug several test pits or sondages in different parts of the cave. In the rear of the cave he had good fortune when he encountered cave bear bones. As winter was near, he only dug a shallow trench, 4 m^3 in size. He carried on excavations the following summer, in 1927, and ended his excavations in September 1928.

On 1 August 1928 during his mountaineering in the Kamnik-Savinja Alps with his friends, S. Brodar also climbed the peak of Olševa. Descending from the top the path took him past the entrance of the cave Potočka zijalka which he then examined. Brodar, a teacher of biology and geology, immediately assumed that this cave could have offered shelter to many mammals escaping the Diluvium and maybe even humans. In the valley Brodar learnt from local people that a student of medicine, Grosz, held a complete skeleton restored in an anatomical position at his home in Bela.

To prevent further illegal excavations by Grosz in the cave on the Yugoslav side of the border (it was then the Kingdom of Yugoslavia), Brodar started his own excavations with financial support from Celje's community and the Museum of Celje. He started to excavate at the rear of the cave where Grosz had been successful. Excavations lasted from 16 September till 6 October 1928. On 21 September Brodar found the first bone points (Fig. 2) and knew this was a Palaeolithic site. In the next few years Brodar also excavated at the entrance of the cave. The excavations continued until 1935.

In the rear of the cave 80 bone points were unearthed from Layers 4 and 5 according to Brodar. Although Brodar found only one small hearth in Layer 4 many small pieces and thin

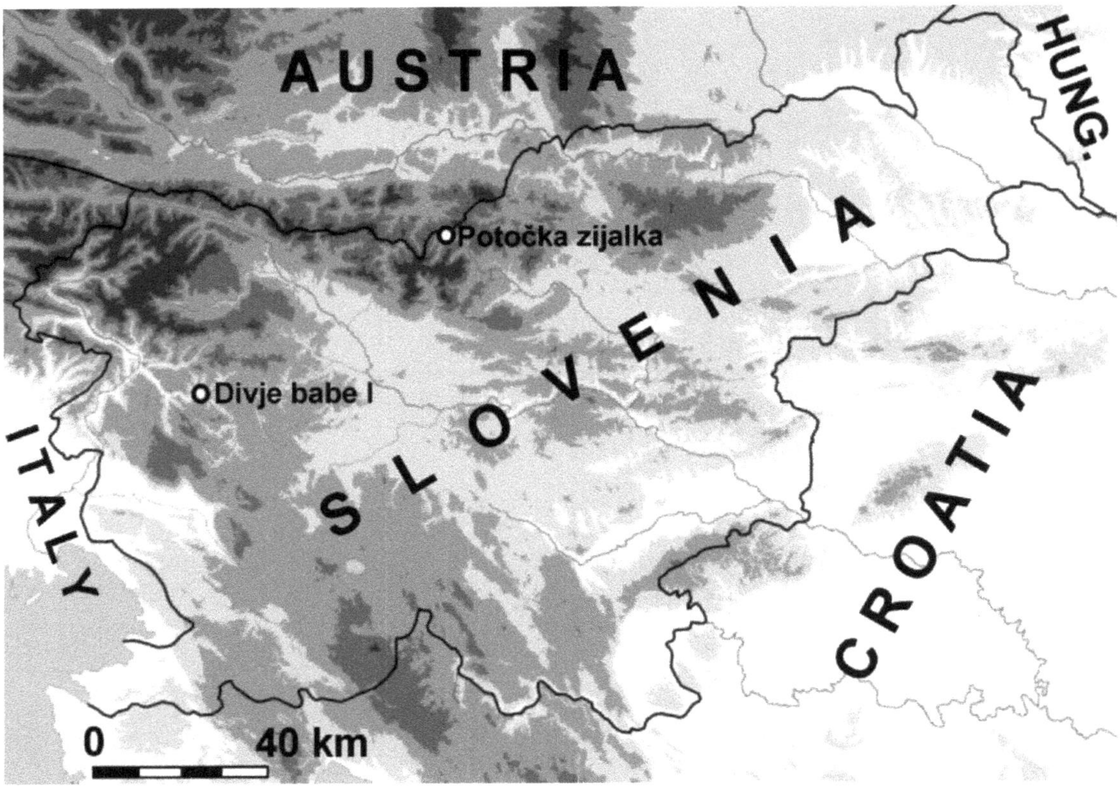

Fig. 1: Slovenia with the two Palaeolitic cave sites and neighbouring countries. White = Lowland, dakest grey = High Alpine mountains.

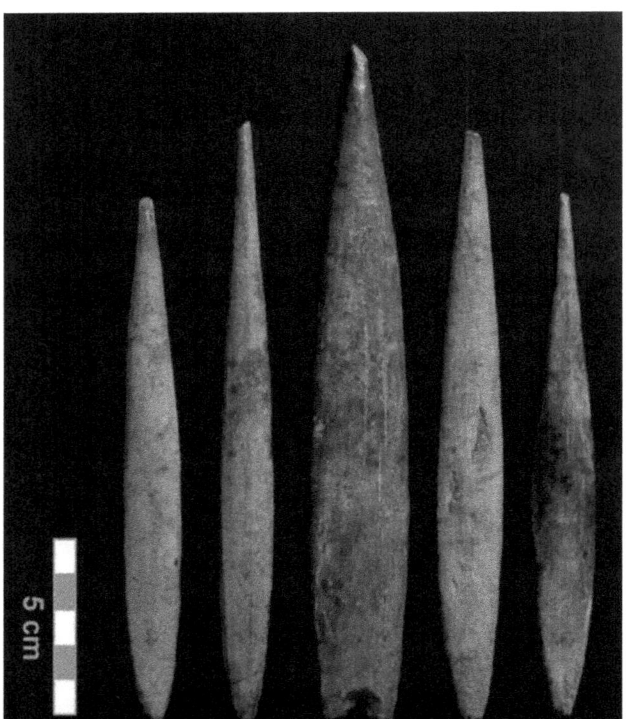

Fig. 2: Some bone points from Potočka zijalka. The biggest bone point is 19 cm long, the smallest only 4 cm.

layers of charcoal were detected all over both layers. In my opinion these charcoal pieces indicate the use of torches in full darkness. Brodar also excavated huge amounts of cave bear bones. On two jawbones (*ramus mandibulae*) with ends removed, he noticed the unusual position of holes above a nerve channel (Fig. 3). He was convinced that the holes are of artificial origin and therefore made by humans. In the rear of the cave the oldest needle of tubular type was also found. He failed to find any stone tools in the darkness at the end of the cave or any other traces of human activities.

At the cave entrance the situation was totally different. In particular, Layer 7 was rich with fireplaces. Brodar detected 17. Around these hearths the concentration of bone points and stone artefacts was highest. 31 bone points in different state of preservation originate from Layer 7. A huge amount of charcoal was found in the most intensively used part of the cave where hearths B4, B5 and B7 were placed. These yielded more than 2 m^3 of pure charcoal, many pieces of which were bigger than a human fist. Layer 8 with sparse traces of human activities represents the oldest occupation layer. Only one small hearth was found in the centre of the cave entrance. In Layer 5 three hearths were at the cave entrance. 11 bone points were found a little deeper in the cave. The youngest traces in the cave left palaeolithic people behind during formation of Layer 3. Only two bone artefacts and two stone flakes were found near the cave entrance. In occupation Layers 5, 7 and 8 at the cave entrance many pieces of red pigment were found.

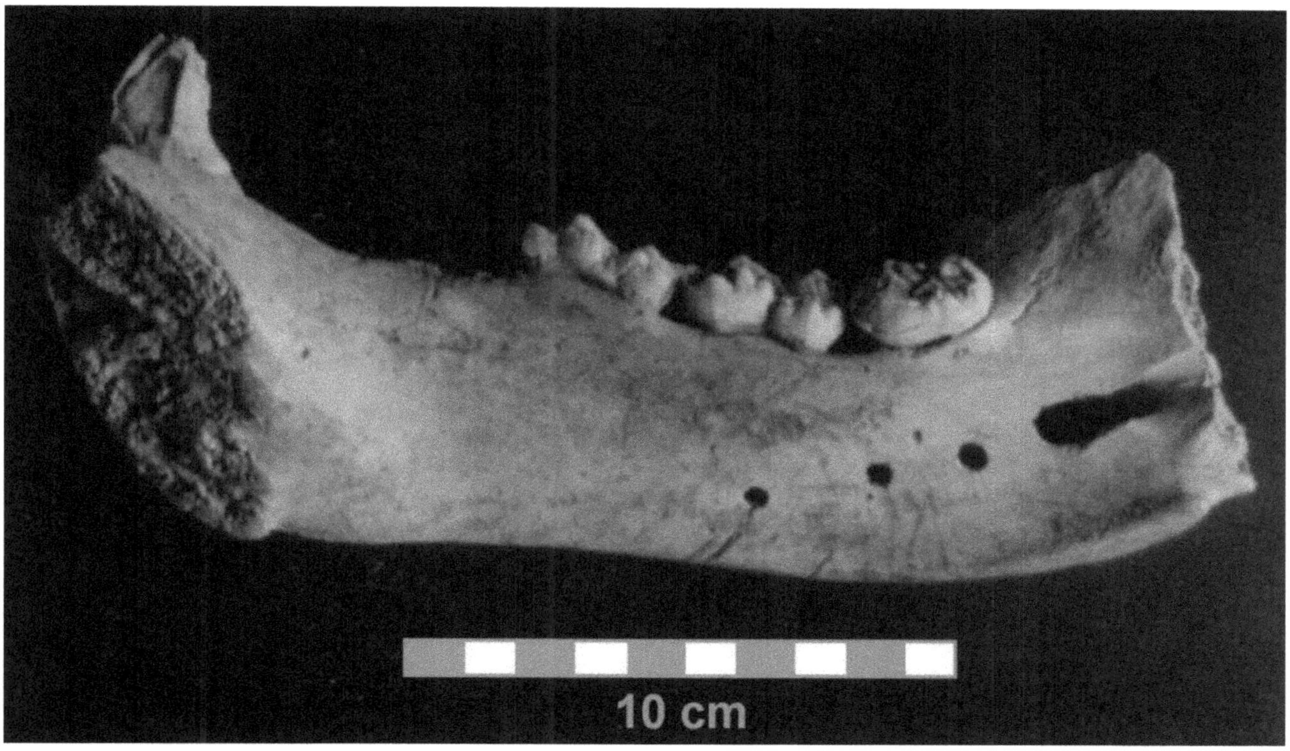

Fig. 3: Several cave bear mandibules with holes above mandibular channel were found in Potočka zijalka. They were used as a special type of flutes.

The new excavation campaign took place between 1997 and 2000 (Pacher *et al.*, 2004). Six bone points were directly dated by the radiometric method AMS ^{14}C. They gave an uncalibrated age of around 30,000 years (Pacher 2001; Rabeder and Pohar 2004).

Most interesting is the fact that over 99 % of all animal bone remains in Potočka zijalka are derived from cave bear (Brodar and Brodar 1983, 90-94). Only small amount of cut marks on cave bear bones made by stone tools suggest manipulation by man (Pohar 2004). Any traces of bone point production in the cave are totally missing. There are also no traces of animals that were hunted by human hunters.

In total, 127 bone points originate from Potočka zijalka. This collection offers a complex view of how they were made and how they were used (Odar 2008). The biggest one is 19 cm long; the smallest only 4 cm. Bone points are believed to be a substantial part of a composite hunting weapon. They were attached to the top of lances, javelins and arrowheads. The hunters who were visiting Potočka zijalka 30,000 years ago were equipped with lances and bows (Odar 2008a, 180-229; Odar 2008b).

With the exception of two bone points, all other bone points from Potočka zijalka were made from cave-bear bones. I have mentioned two cave-bear jawbones with an unusual series of holes. Several more such jawbones were found in occupation layers at the cave entrance (Brodar and Brodar 1983, 155-158). They were thought to be a certain type of flute. Experiments with replicas clearly show that such flutes have a wide range of sounds which are easily transformed into music (Omerzel-Terlep 1996). One jawbone of excellent preservation was perforated at its end. Probably this jawbone once hung on string.

The fact that the bone points and flutes were made from cave bear bones shows a special relationship of people to cave bears. Indigenous people from Alaska and other people living around the circumpolar north once believed that bears are close relatives and that people are able to have common successors with bears (Binford 1997; Ingold 2003, 114-115). Bears are the biggest beasts with outsize powers in the northern hemisphere. That is the reason why people attached points to their armour that were made from bear bones. Such armour was believed to possess the powers of bears.

Ethnographic sources reveal that people rarely decided to hunt bears. When they did, the hunt took place during the hibernation period. A bear awakened during this period was much less dangerous. The chosen bears did not weigh more then 200 kg. It was too dangerous to hunt bigger bears. Bear hunting was practised only on special occasions (Pacher 1997; Binford 1997). Today we know that the cave bears encountered natural death in caves during the hibernation period due to insufficient food supply. The average weight of cave bears was much bigger then the weight of still living bears. Adult cave bear males weighted over 500 kg, females around 300 kg.

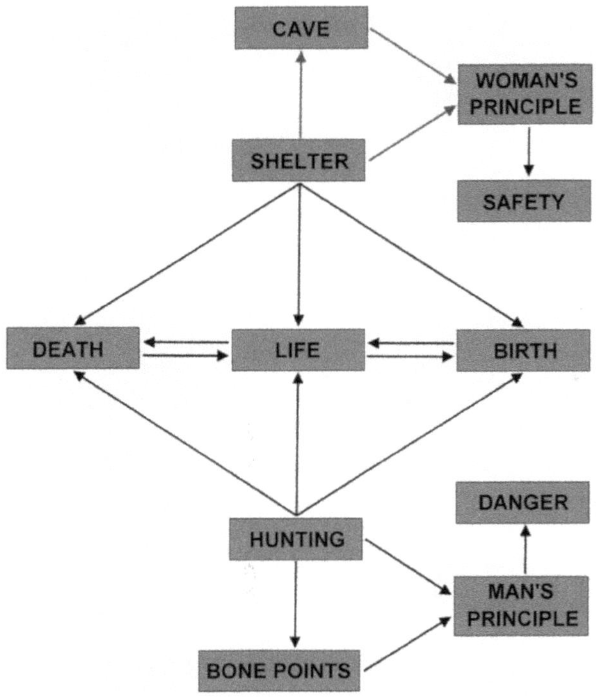

Fig. 4: Bone points in the cave show perpetuation of life cycle on symbolic level.

Hunting plays an important role in the animistic world. It provides a circulation of powers between people and animals. Many indigenous people all over the world explain hunting as an act of sexual interaction between hunter and prey. Piercing the prey with hunting weapons and the subsequent bleeding is compared to the first sexual experience of a girl with a man (Ingold 2003, 13, 113-115). This could explain the presence of red pigment at the cave entrance of Potočka zijalka. Hunting does not mean exclusion from the cycle of life; rather it is an act of renewal. Animals are deemed to offer their body and powers to people. On the other hand people take care that the animal remains are handled with respect (Kurtén 1976, 91; Zachrisson and Iregren 1974.

In my opinion hunting and shelter could mean two general principles, of man and of woman (Fig. 4). Both enable the life cycle from birth until death. Visitors left bone points or hunting armoury intentionally in Potočka zijalka. The whole context shows the perpetuation of the life cycle at a symbolic level. Cave bears were born in the caves, they lived in caves, and they died therein.

Bone points in the cave may represent sexual intercourse between a god and goddess. The east-west orientation of the Olševa Mountain offers an answer to the question of the identity of these divinities (Fig. 5). In the morning the Earth gives birth to the Sun. In the evening the Sun goes down

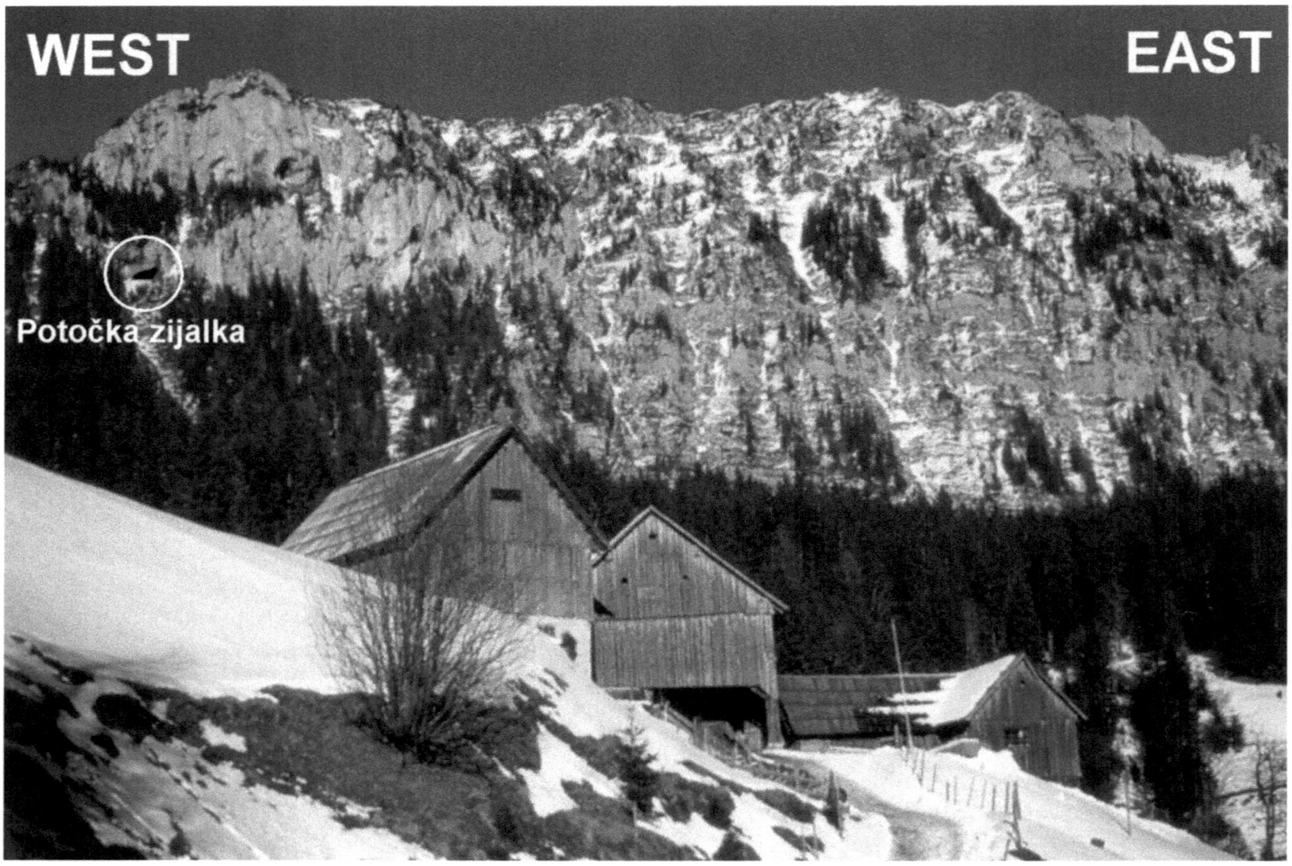

Fig. 5: Potočka zijalka is positioned on Olševa Mountain 1630 m above sea level

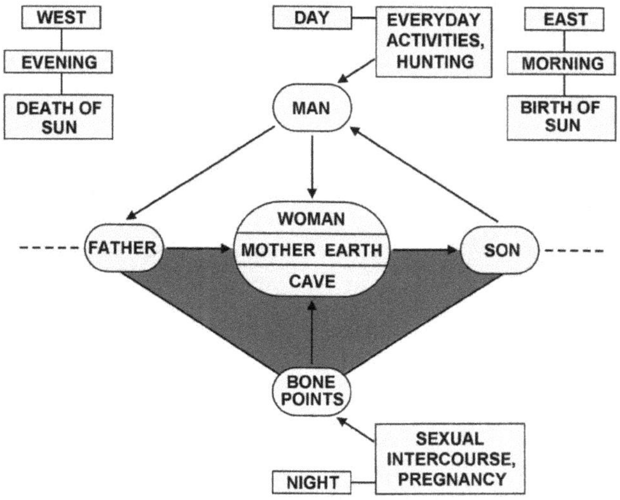

Fig. 6: A geo-solar cosmological principle in the case of Potočka zijalka

and begins an intimate relation with the Earth as an adult male. This permits the rebirth of the Sun on the following day (Fig. 6). A position in front of the cave entrance gives one the impression of standing at the top of the world. This could be clear evidence of the incorporation of animism into a geo-solar cosmological principle in Potočka zijalka.

The finds at Divje babe I and symbolic interpretation

The Divje babe I cave site (Gauss Kruger co-ordinates x = 108160,75m, y = 416555,77m) in the Idrijca valley in western Slovenia lies at an altitude of 448m, beneath the edge of the Šebrelje plateau (600-700m above sea level). Šebrelje is part of the foothills in the most south-easterly part of the Alps. The cave entrance is orientated to the north and being hidden in woodland it is difficult to see it from the Idrija river valley (Figs 1, 7).

The first systematic excavations by Mitja Brodar were at the cave entrance in 1978 (test trench) and from 1980 to 1986. He method used a combination of horizontal and vertical excavation. After partly removing sediments, he documented the sediments in profiles. Finds were documented within all three coordinates. Due to a high variability and disturbances in sedimentation, a common idealized profile was in use to describe all sediments. With the exception of artefacts and some probes, other data were not quantified. From this period of excavations we have only qualitative bases of data or descriptions. No radiometric analyses were done and sediments were not sieved (Turk 2007, 16).

Ivan Turk began new excavations in 1989 which continued until 1999. He did not follow the wish of the previous excavator and decided to excavate in the inner part of the

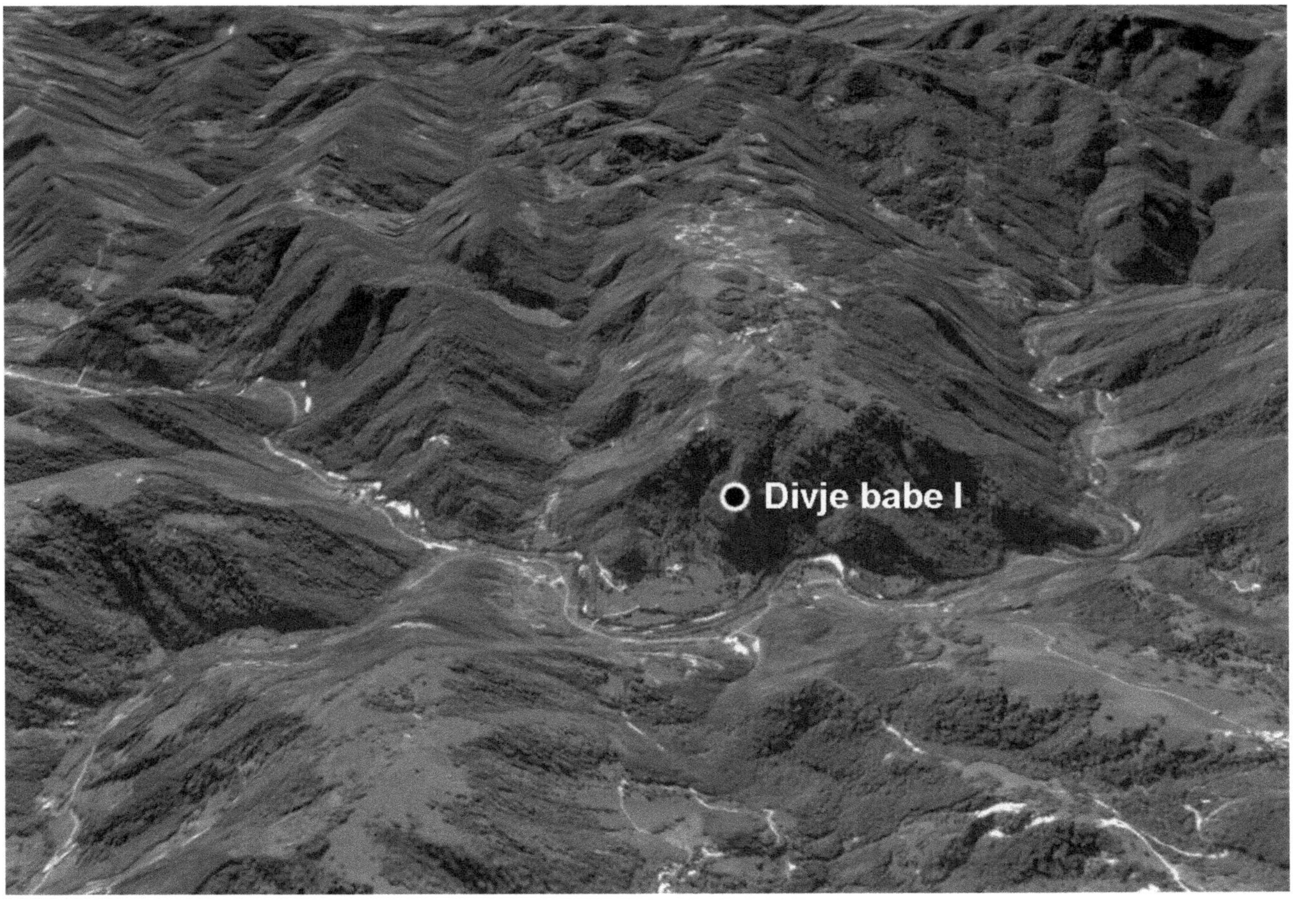

Fig. 7: Divje babe I are positioned on the northern slope of the Šebrelje plateau

cave with improved methods. All sediments were excavated within a one square-metre grid and 12cm thick slides. All excavated slides, without exception, were water sieved on three sieving grids with different hole sizes (10mm, 3mm and 0.5mm). This method enabled Ivan Turk to control all the changes in sedimentation within the excavated area of the cave. More than 80 ESR and AMS ^{14}C dates for all sedimentation levels were done. They showed a time span between 115,000 and 40,000 years before present (Turk 2007).

Rare but important finds from Divje babe I indicate that people vsited the cave only occasionally. Fragments of bone points were found. The earliest is dated to 115,000 years, whilst the latest dates to 40,000 years (Turk *et al.* 2001; Turk 2007, 159-165). However the most interesting is large hearth in Layer 8. The Neanderthal flute, 60,000 years old, was found there (Horusitzky 2003; Turk 1997; Turk *et al.* 2001, 2003, 2005). In close proximity two cave bear skulls were found together facing away from each other. The male skull was orientated to the west, the female to the east. Both skulls were turned around. The male skull has unusual symmetrical holes in both massive spurs behind the mandibular articular depression (*fovea articularis mandibulae*). The two holes were probably made for the purpose of strongly attaching or hanging the skull (Turk 2003; Turk 2007, 338).

The significant loss of cave bear canines in the occupation layers in this cave was noticed and the much higher fragmentation of cave bear skulls (Turk 2007, 337-339). Divje babe I did not yield many skeletal remains of meat-bearing animals that would have been of interest to human hunters. Intentional fragmented cave-bear skulls and a significant absence of cave-bear canines in the occupation layers may indicate an animistic relationship of people to cave bears. The famous Neanderthal flute was made of cave bear bone (Fig. 8).

But what is the real context of the hearth or fireplace with the flute and the two cave-bear skulls nearby? The secret is to be found in the old Slavic pagan name of the northern slope of Šebrelje plateau. The toponym 'Divje Babe' means 'Savage Women'. An incredible sight can be seen if one stands on the other side of the Idrija valley and looks towards Divje babe. A natural rock formation in the form of a large vulva is visible directly below the cave (Fig. 9)—and this is exactly what Neanderthals probably saw. The two cave-bear skulls are the most complete skulls in the cave. The female skull was orientated to the east, where the Earth gives birth to the Sun, while the male skull was orientated to the west, where the Sun goes down. These two skulls show exactly the same geo-solar cosmological pattern as that previously defined in Potočka zijalka. The hearth, the two intentionally-orientated cave-bear skulls and the flute were positioned in the womb. The whole rock formation of the Divje Babe slope, together with the cave entrance, is visible only in winter when the trees are leafless. The cave was not appropriate for permanent living. It is almost

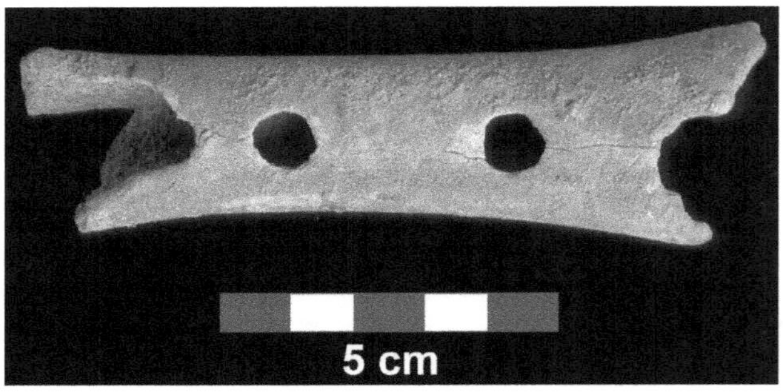

Fig. 8: The Neanderthal flute came from layer 8, which was dated to 60 000 years.

without sunlight and access is tricky, even dangerous due to the steep slope. The rarity of stone tools and fragments of bone points confirm this supposition.

One cosmology, different manifestations

(a) Palaeolithic

The most important region with Aurignacian rock art depictions in Europe is certainly the Dordogne in France. There, many blocks of limestone with engraved pictograms are to be found. One of the best known sites with Aurignacian art is Le Grand Abri La Ferrassie near Les Eyzies-de-Tayac. In sediments from the Lower Aurignacian more than 20 engraved blocks from rock fall were found. In most cases sexual symbols in the form of *vulvas* and *phalli* were depicted on them. Similar depictions from the Aurignacian are known from Abri Castanet, Abri Blanchard, Abri Cellier in Tursac and from the Chauvet and Cosquer caves (Clottes 2001; Clottes, Lewis-Williams 1996; Floss 2009; Le Guillou 2001).

Here one should mention the oldest known Venus figurine from Hohle Fels, South Germany (Conard 2009a). The ivory figurine was found in 2008 at the base of the Aurignacian layer that was dated by the radiometric method AMS 14C to an age between 31,000 and 40,000 years ago. Until this discovery there was an assumption that the tradition of making figurines denoting sexual parts of the body first appeared in the Gravettian. In the same year, 2008, a second surprising find from the lower Aurignacian layer came to light. A flute made of bird bone was found in the same horizon as the Venus figurine only 0.7m away (Conard 2009b). In my opinion the composition of these artefacts indicate some ritual practice in Hohle Fels 30,000–40,000 years ago. Three flutes were also found in an Aurignacian layer of Geiβenklösterle, and are 32,000 to 36,800 years old (Münzel and Conard 2009).

As in the case of Potočka zijalka, bone points played a crucial role in the complex cosmological system of Palaeolithic hunters visiting this cave. Several beautifully ornamented bone points were found in the French cave Lascaux (Allain 1979). In many cases simple projectiles were depicted over different animal species as in Niuax

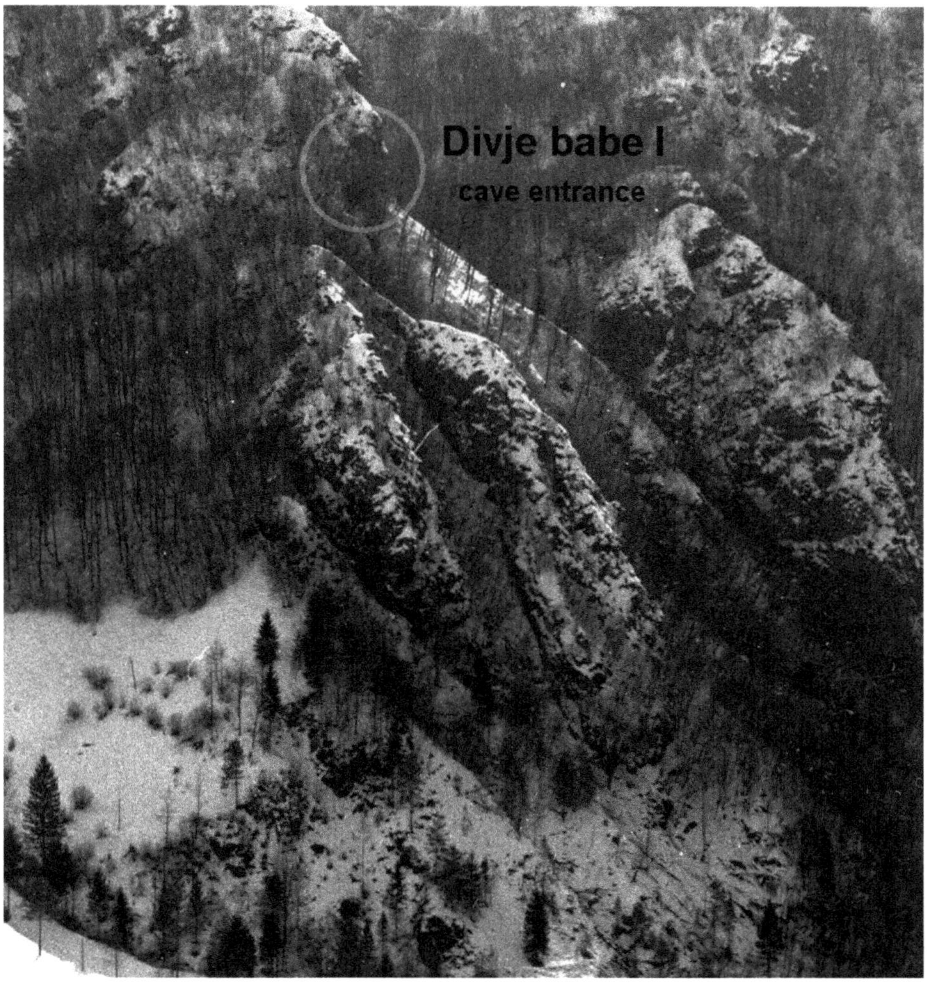

Fig. 9: A natural rock formation in the form of a large vulva below the cave entrance

or Lascaux for example (Clottes 1997; Leroi-Gourhan and Allain 1979). Usually such depictions are interpreted as hunting scenes. Clottes and Lewis-Williams (1996) have connected cave rock art with trance and magic that occurs during shamanistic practice in decorated caves. Shamanism is firmly connected with dance and music, as are many other rituals. We have clear evidence of music instruments in the caves. Divje babe I, Potočka zijalka (Slovenia), Hohle Fels, Geiβenklösterle (Germany) and the much younger Isturiz (France) are only a few of the examples.

I am sure that Upper Palaeolithic cave rock art was a substantial part of rituals in everyday life. The presence of bone points in the caves, whether as objects or depicted as hunting armoury on cave walls, express a desire of the Palaeolithic hunters after renewing the living world. A feature of the big rocky hill protruding from the earth at the French village of Niaux gives the impression that Mother Earth is close to giving birth. Depicted animals together with significant human remains inside the caves were actually positioned in the womb. Animals pierced with hunting armour (lance, javelin or arrow) were probably chosen as totemic animals by the Palaeolithic communities. In the Alpine region the largest and mightiest animal was certainly the cave bear. It is no wonder, then, that Palaeolithic hunters once inhabiting the Alps chose this species for their totemic animal as in the case of Divje babe I (Mousterian) and Potočka zijalka (Aurignacian).

With the ending of the Ice Age the environment changed dramatically all over Europe. Many animal species became extinct, and many others moved north to colder regions to be near the surviving ice cover. Hunters must have adapted to new living conditions, but the substantial parts of symbolism that had evolved in the Palaeolithicl were not forgotten. They were transmitted into Mesolithic and later periods.

(b) Mesolithic / Neolithic

Maybe the most intriguing is the appearance of megalithic tombs on the Iberian Peninsula (Leisner and Leisner 1951, 1956, 1959 and 1965). The first such tombs are from the transition period from Mesolithic to Neolithic 7,000 years ago. Instead of using natural caves, people started to build artificial caves within the landscape for their ritual practices. Through time these megalithic constructions became more complex. The path to the chambers inside the tomb leads through narrow passages. Without exception the entrance and passageway were orientated between south and east. The shapes of the megalithic structures resemble vulvas (passages) connected with wombs (chambers). In most tombs ritual objects in the shape of phalli or beautifully-shaped small stone arrowheads were found. This clearly expresses on a symbolic level the people's wish after death. It is important to mention that, at least with the Mesolithic, bows and arrows became the most important long-distance weapon. This is why bows and arrows are found in many graves from all succeeding periods until the end of the Early Middle Ages all over the Europe (Odar 2006).

(c) From Hellenism to Christianity

One of the most famous Greek tragedies is the story of Oedipus who was the son of Laius and Jocasta, king and queen of Thebes. After being married for some time without children, his parents consulted the Oracle of Apollo at Delphi about their childlessness. The Oracle prophesied that if Laius should have a son, the son would kill him and marry Jocasta. Fearing the prophecy, Laius pierces Oedipus' feet and leaves him out to die, but a herdsman finds him and takes him away from Thebes. Little Oedipus came to the house of Polybus, king of Corinth, and his queen, Merope, who were without children of their own. Many years later, Oedipus is told that Polybus is not his real father but when he asks his parents, they deny it. Oedipus, uncertain, seeks counsel from the same Delphic Oracle. The Oracle does not tell him the identity of his true parents but instead tells him that he is destined to kill his father and marry his mother. In his attempt to avoid the fate predicted by the Oracle, he decides to flee from Corinth to Thebes. Laius, meanwhile, ventures out to find a solution to the Sphinx's riddle. As prophesized, Oedipus crossed paths with Laius and this leads to a fight where Oedipus slays Laius. Oedipus then defeats the Sphinx by solving a mysterious riddle to become king of Thebes. He marries the widow queen Jocasta not knowing she is his true mother. A plague falls on the people of Thebes. Upon discovery of the truth, Oedipus blinds himself and Jocasta hangs herself. After Oedipus is no longer king, Oedipus' sons kill each other. Left behind is Oedipus's only daughter Antigone.

In one Slavic grave from Slovenia an interesting brooch was found. On the small round brooch is a depiction of *Centaur-Sagittarius* with very long hair. The unusual composition of the whole context led me to conclude that the transfer of antique iconography into Slavic mythology and its main myth had occurred (Odar 2003, 91-93). In the main Slavic myth the storm god Perun lives in the Heavens. God Veles lives in the Underworld as a chthonic being. Perun and Veles are fighting for the goddess Mokoš. Mokoš, who is a wife of Perun, actually stands for Mother Earth. While fighting, Perun is throwing thunder claps against Veles with a bow. The possibility of transforming into a human being or into a horse gives Veles a chance to hide himself from the anger of Perun. In the endless fight for Mokoš both gods are the winners with changing luck. Perun is the winner of the day and the light (spring, summer), and Veles is the winner of the night and the darkness (autumn, winter) (Fig. 10). This iconography of *Centaur-Sagittarius* in the Slavic grave thus represents the trinity of the main Slavic gods, Perun (armed with a bow), Veles (as a horse) and Mokoš (with long hair). Here I have to add that the Slavic god Perun is of Indo-European origin. By Kashubs Thursday is

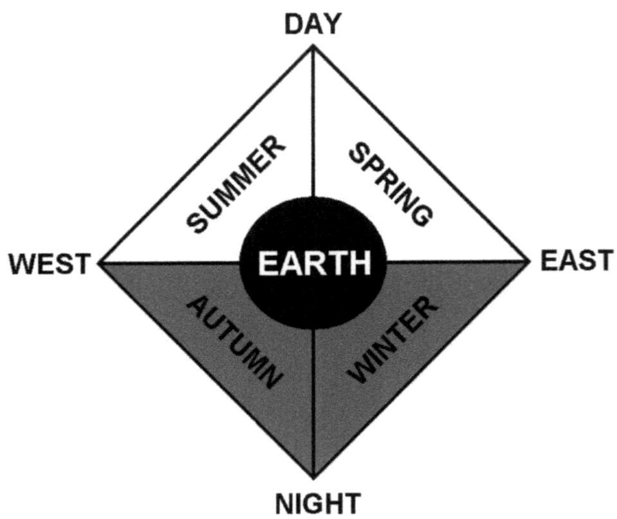

Fig. 10: Geo-solar cosmological principle grounds on cyclical rotation of the Sun around the Earth. Compare with Fig. 6 in this text.

named Peraunědan (Perun's day), by Germans Donnerstag (Thunder's day), by Italians giovedi.

The trinity, as best known from Christianity today, has very deep roots in human history. Those roots most probably began to grow already in the Middle Palaeolithic (Fig. 11). If this is true, then it is no wonder that a church of Holy Spirit was erased beneath Potočka zijalka at foothill of Olševa, directly in the middle of the *axis mundi*, the sunrise and the sunset. I do not need to mention the important role of Holy Spirit in Christianity.

Conclusions

The basic idea of trinity is the circulation of the Sun around the Earth. With the development of consciousness, people became aware of the powers of the Sun and its importance for their existence and the existence of the rest of the living world. The most impressive moments of the day are certainly sunrises and sunsets. Even today we are impressed by these natural phenomena. The impression is that Mother Earth gives birth to the Sun in the morning, and the Sun begins an intimate relationship with Mother Earth in the evening, and this may have led Palaeolithic people to evolve a geo-solar cosmology. This might have happened already in the Middle Palaeolithic in Europe. In the Alps people encountered the mightiest animal of the region, the cave bear. Because of its immense power the cave bear had already become a totemic animal by Neanderthals. When anatomically-modern humans arrived in Europe, they shared living space with Neanderthals for at least 10,000 years. During this time-span some habits evolved by Neanderthals were taken over by the newcomers. Among those habits we detect a possible geo-solar cosmology and a special attitude for bears. Both were evolved further as this article has tried to show.

geo-solar cosmology	TRINITY		
orientation	WEST	EARTH	EAST
Divje babe I cave	male bear skull looking to the west	Divje babe hill slope	female bear skull looking to the east
Potočka zijalka cave	Evening Sun	Olševa mountain	Morning Sun
Greek tragedy	Anacreon	Iokastis	Oedipus
Main Slavic myth	Perun	Mokoš	Veles
Christianity	Holy Father	Saint Mary and/or Holy Ghost	Holy Son

Fig. 11: Some cases from different periods of human history are showing different manifestation of geo-solar cosmology.

References

ALLAIN, J. (1979). L'industrie lithique et osseuse de Lascaux. In Leroi-Gourhan A. and Allain J. (eds.), *Lascaux inconnu,* 87-120. Paris, Éditions du CNRS.

BINFORD, L. (1997). Linking Ethnografic Information on Man-Bear Interaction to European Cave Bear Deposits, in TILLET, T. and BINFORD, L. (eds.), *Man and Bear.* International meeting, November 4-6, 1-44. Auberives.

BRODAR, S. and BRODAR, M. (1983). *Potočka zijalka: Visokoalpska postaja aurignacienskih lovcev.* Ljubljana, SAZU.

CLOTTES, J. (1997). *Niaux: Die altsteinzeitlichen Bilderhöhlen in der Ariège.* Sigmaringen, Thorbecke.

CLOTTES, J. (ed.) 2001. *La grotte Chauvet: L'art des origines.* Paris, Seuil.

CLOTTES, J., and LEWIS-**WILLIAMS,** J.D. **(1996).** *Les Chamanes de la Préhistoire: Trans et Magie dans les Grottes Ornées.* Paris, Editions du Seuil.

CONARD, N. J. (2009a). Die erste Venus. Zu ältesten Frauendarstellungen der Welt, in S. RAU, D. NAUMANN and M. BARTH (eds.), *Eiszeit: Kunst und Kultur,* 268-271. Ostfildern, Thorbecke.

CONARD, N. J. (2009b). Die Anfänge der Musik. Eine Knochenflöte aus dem unteren Aurignacien, in S. RAU, D. NAUMANN and M. BARTH (eds.), *Eiszeit. Kunst und Kultur*, 324-326. Ostfildern, Thorbecke.

FLOSS, H. (2009). Kunst schafft Identität. Das Aurignacien und die Zeit der ersten Kunst, in S. RAU, D. NAUMANN and M. BARTH (eds.), *Eiszeit: Kunst und Kultur,* 248-258. Ostfildern, Thorbecke.

LE GUILLOU, Y. (2001). Les representations humaines. In J. CLOTTES (ed.), *La grotte Chauvet: L'art des origines,* 167-171. Paris, Seuil.

HORUSITZKY, F. Z. (2003). Les flutes paleolithiques: Divje babe I, Istallosko, Lokve etc. Point de vue des expert et des conestataires. *Arheološki vestnik* 54, 45-66.

INGOLD, T. (2003). *The Perception of the Environment: Essays in livelihood, dwelling and skill.* London - New York, Routledge.

KUNEJ, D. (1997). Acoustic findings on the basis of reconstructions of a presumed bone flute. In TURK, I. (ed.), *Mousterian bone flute and other finds from Divje babe I cave site in Slovenia,* 185-197. Ljubljana, Založba ZRC.

KUNEJ, D. and TURK, I. (2000). New Perspectives on the Beginnings of Music: Archeological and Musicological Analysis of a Middle Paleolithic Bone 'Flute', in N. L. WALLIN, B. MERKER, S. BROWN (eds.), *The Origin of Music,* 235-268. Cambridge (Massachussetts), London (England), MIT Press.

KURTEN, B. (1976). *The Cave Bear Story: Life and Death of a Vanished Animal.* New York, Colmbia University Press.

LEISNER, G. and LEISNER, V. (1951). *As antas do Conchelho de Reguengos de Monsaraz.* Lisboa, Instituto de Alta Cultura.

LEISNER, G. and LEISNER, V. (1956). *Die Megalithgräber der Iberischen Halbinsel.* Berlin, Madrider Forschungen Band 1.

LEISNER, G. and LEISNER, V. (1959). *Die Megalithgräber der Iberischen Halbinsel.* Berlin, Madrider Forschungen Band 2.

LEISNER, G. and LEISNER, V. (1965). *Die Megalithgräber der Iberischen Halbinsel.* Berlin, Madrider Forschungen Band 3.

LEROI-GOURHAN, A. and ALLAIN, J. (eds.) (1979). *Lascaux inconnu.* Paris, Éditions du CNRS.

MUNZEL, S., and CONARD, N. J. (2009). Flötenklang aus fernen Zeiten, in S. RAU, D. NAUMANN and M. BARTH (eds.), *Eiszeit. Kunst und Kultur,* 317-321. Ostfildern, Thorbecke.

ODAR, B. (2003). *Lokostrelstvo v pozni antiki in zgodnjem srednjem veku (Archery in Early Middle Ages and Late Antiquity).* Unpublished MA thesis, University of Ljubljana.

ODAR, B. (2006). The archer from Carnium. *Arheološki vestnik* 57, 243-275.

ODAR, B. (2008). *Izdelava in uporaba koščenih konic iz Potočke zijalke (Making and use of bone points from Potočka zijalka).* Unpublished PhD thesis, University of Ljubljana.

ODAR, B. (2008). A Dufour bladelet from Potočka zijalka. Arheološki vestnik 61, in press.

ODAR, B. (2010). Archers in Potočka zijalka?. *Arheološki vestnik* 61, in press.

OMERZEL-TERLEP, M. (1996). Koščene piščali. Pričetek slovenske, evropske in svetovne inštrumentalne glasbene zgodovine. *Etnolog* 6 (LVII), 235-291.

OMERZEL-TERLEP, M. (1997). A typology of bone whistles, pipes and flutes and presumed palaeolithic wind instuments in Slovenia. In TURK, I. (ed.), *Mousterian bone flute and other finds from Divje babe I cave site in Slovenia,* 185-197. Ljubljana, Založba ZRC.

PACHER, M. (1997). *Der Höhlenbärenkult aus ethnologischer Sicht.* Wissenschaftliche Mitteilungen des Niederösterreichischen Landesmuseum 10, 251-375.

PACHER, M. (2001). *New excavation campaigns in the Upper Pleistocene cave bear site Potočka zijalka, Slovenia - state of investigation.* Cadernos do Laboratorio xeolóxico de **Laxe** 26, 301-310.

PACHER, M., POHAR, V. and RABEDER, G. (eds.) (2004). *Potočka zijalka. Paleontological and Archaeological results of the Campaigns 1997-2000.* Bad Vöslau, Österreichische Akademie der Wissenschaften.

POHAR, V. (2004). Stone- and Bone Artefacts from the Excavation 1997-2000 in Potočka zijalka (Slovenia), in M. PACHER, V. POHAR, G. RABEDER (eds.), *Potočka zijalka. Paleontological and Archaeological results of the Campaigns 1997-2000,* 211-215. Bad Vöslau, Österreichische Akademie der Wissenschaften.

RABEDER, G. and POHAR, V. (2004). Stratigraphy and Chronology of the Cave sediments from Potočka zijalka (Slovenia), in M. PACHER, V. POHAR, G. RABEDER (eds.), *Potočka zijalka. Paleontological and Archaeological results of the Campaigns 1997-2000,* 235-245. Bad Vöslau, Österreichische Akademie der Wissenschaften.

TURK, I. (ed.) (1997). Mousterian bone flute and other finds from Divje babe I cave site in Slovenia. Ljubljana, Založba ZRC.

TURK, I., DIRJEC, J., BASTIANI, G., PFLAUM, M., LAUKO, T., CIMERMAN, F., KOSEL, F., GRUM, J. and CEVE, P. (2001). New analyzes of the ''flute'' from Divje babe I (Slovenija). *Arheološki vestnik* 52, 25-79.

TURK, I., BASTIANI, G., BLACKWELL, B. A. B. and HORUSITZK, Z. (2003). Putative Mousterian bone flute from Divje babe I (Slovenia): pseudoartefact or true flute, or who made the holes. *Arheološki vestnik* 54, 67-72.

TURK, I. (2003). Humans and Carnivores In Slovenia durinng Upper Pleistocene. Neanderthals and the Cave Bear, in B. KRYSTUFEK, B. FLAJS MAN and H. I. GRIFFITHS (eds.), *Living with Bears. A large European Carnivore in a Shrinking World,* 43-57. Ljubljana, Ecological Forum of the Liberal Democracy of Slovenia.

TURK, I., PFLAUM, M. and PEKAROVIC, D. (2005). Results of computer tomography of the oldest suspected flute from Divje babe I, Slovenia: Contribution to the theory of making holes in bones. *Arheološki vestnik* 56, 9-36.

TURK, I. (ed.) (2007). *Divje babe I. Part 1.* Ljubljana, Založba ZRC.

ZACHRISSON, I. and IREGREN, E. (1974.) *Lappisch bear graves in northern Sweden: An Archaeological and Osteological Study.* Stockholm, Kungliga Vitterhets Historie och Antikvitets Akademien.

Sacred Sites and Symbolism in the Neolithic Landscape of Barbagia di Seulo, Central Sardinia

M. Giuseppina Gradoli

COMET – Valorizzazione Risorse Territoriali and ISSEP Sardinia, Via dei Colombi, 31, 09126 Cagliari

G. Terence Meaden

Oxford University, Department of Continuing Education (Archaeology) and Kellogg College, 62 Banbury Road, Oxford, England, OX2 6PN

***Abstract**: Sacred ritual sites, some of them Neolithic or Bronze Age, are identified for the first time in the Barbagia di Seulo region of mountainous Central Sardinia. They include rockshelters, caves and a natural rock tunnel. Some discoveries resulted from place-name studies and knowledge of traditional folklore. Important was the recognition of symbolic concepts as previously known to be typical of Neolithic fertility and Mother Earth symbolism elsewhere in Europe, indeed worldwide. It is suggested that such similarities may be owed to fundamentals in cultures and in the psyche that are common to farming communities everywhere through the people's devotion to their ancestors and need for fertility success in their fields, their animals and their women.*

Four sites are introduced in this paper, and two more in further papers. First is a splendid rock-shelter site called 'Su Cannisoni' or 'Su Cunnisoni' near to which is a small bone-filled cave called 'Sa Grutta 'e Is Bittuleris'. The local area, known as 'Sa Omu 'e Is Ossus', means 'The House of the Bones'. A third site is a rockshelter with cascade and pool called 'Su Stampu 'e Su Turrunu', while a fourth site is a rock tunnel 'Is Caddaias' through which passes fast flowing water. It is important that these sites are given the statutory protection that they deserve.

***Key words**: Neolithic, Sardinia, Su Cannisoni, Su Stampu 'e Su Turrunu, Is Caddaias, Mother Earth, symbolism, simulacra, images.*

***Riassunto**: Diversi siti rituali sacri sono stati identificati, per la prima volta, nella regione montuosa della Barbagia di Seulo, Sardegna Centrale. Essi comprendono ripari sotto roccia, grotte e tunnel nella roccia naturale. Importante è stato la studio dei toponimi ed il riconoscimento di concetti legati al simbolismo della fertilità e della Dea Madre, già noti per il Neolitico dell'Europa e del Mondo intero.*

***Parole chiave**: Neolitico; Sardegna; Simbolismo; Religione della fertilità; Madre Terra.*

Introduction

These papers on Sardinian archaeology announce the discovery of an important, previously-unrecognised, ritual landscape in the central heights. The Barbagia di Seulo is a mountainous limestone region cut by wooded valleys including the river Flumendosa and tributaries. A study of place-names using linguistics and knowledge of traditional folklore raised the possibility that prehistoric sites may be found, and subsequent investigation led to the discovery of Neolithic and Bronze Age sites of various character including caves and rock shelters (Gradoli 2005). In some cases concepts appropriate to Neolithic and Bronze Age symbolism elsewhere in Europe were found to be present as topographical and landscape imagery of a type deemed to have fertility and Mother Earth connotations (Gradoli and Meaden 2012).

Four of the sites are introduced in this paper. Investigation produced clear evidence for a former Neolithic/Bronze Age presence at two of them. Two of the four sites are rock shelters, one a bone-filled cave close to one of the shelters, and the fourth a rock tunnel with fast-flowing water. Two other caves with Neolithic or Bronze Age remains or artefacts are described and discussed in the following papers (Gradoli and Meaden 2012a, 2012b).

Sacred landscapes

According to Carmichael et al. (1994, p.3, quoted in Buikstra and Charles, 1999) a "Sacred Landscape carries with it a whole range of rules and regulation regarding people's behaviour in relation to it and implies a set of beliefs often in relation to the spirits of the ancestors, as well as more remote or powerful gods or spirits".

Some remarkable landscape features seem to be common to humankind and individual and social identities (Hawkes 1954,167) while memory and culture act like "carrier waves" transmitting information across time and space, even when a ritual is retained but its meaning lost. Moreover, through rituals, societies link economic

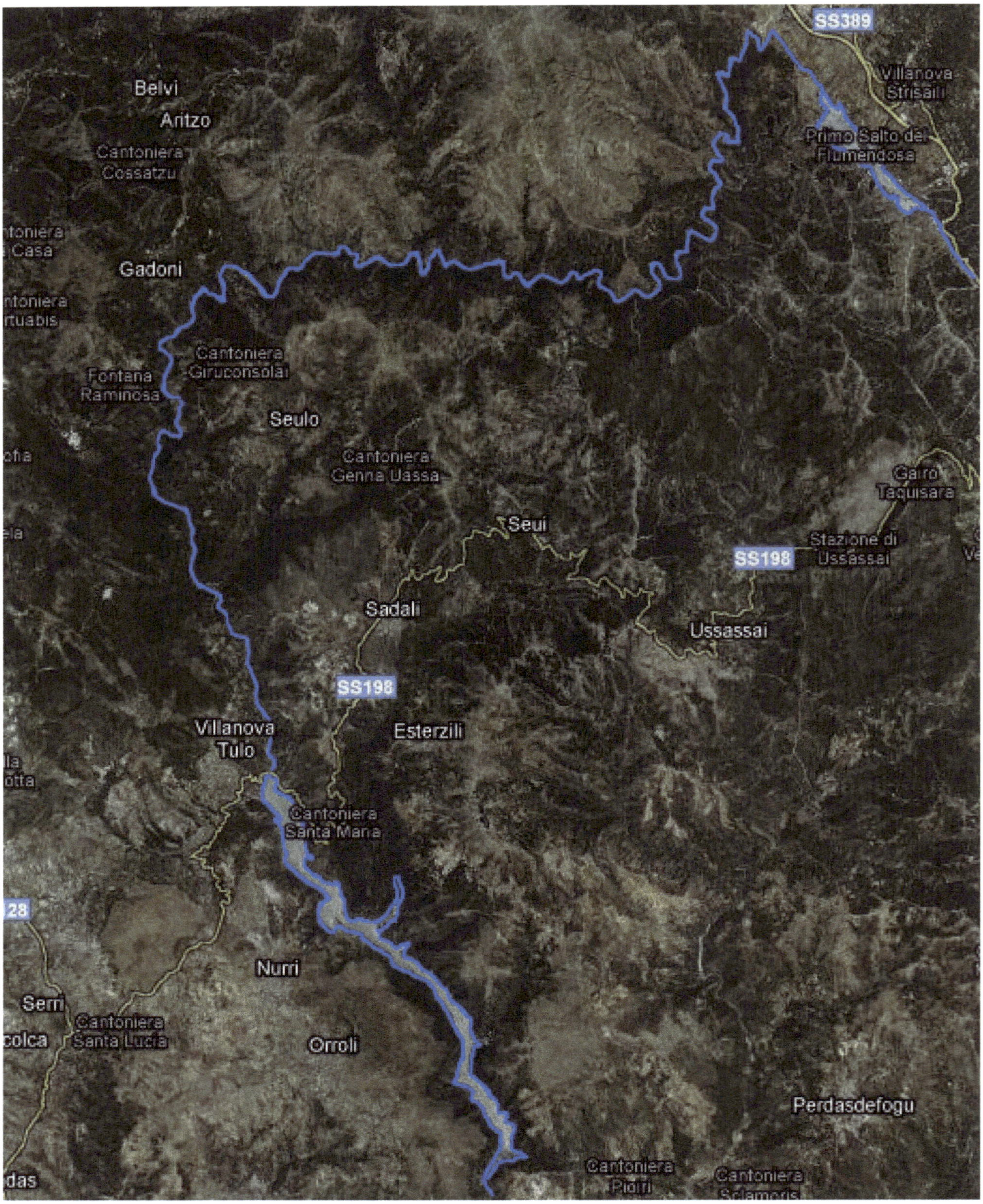

Fig. 1. Showing the general location of Barbagia di Seulo in the south of Central Sardinia. In blue colour is the River Flumendosa that has served as a natural border since Prehistoric times.

success, belief and elite power and employ the strong tugs of individual and group experience to demonstrate their social and religious superiority (Crumley 1999).

Often prominent landmarks, geological formations, mountain ranges, steep valleys, and waterfalls, water-pools and other "liminal places" such as caves were subject to "special attention" and were marked by open-air rock art (Bradley 1991) or were places where offerings were made (McEwan and Guchte 1992) or shrines constructed (Aldenderfen 1990). Bradley (1997: 215) explains how some of these images associated with ceremonial centres extended to natural places of the landscape and, at certain times, became vested with a "supernatural" identity (Ucko 1994). Indeed, when practises continue at the same locations over many generations these places and their surrounding landscapes become ever more symbolically charged—i.e. sacred—and it was believed that only the highest levels of religious knowledge could approach them (Taçon 1999).

The ancient Greeks, Romans, Maya, Chinese, Britons, Vikings and people of south-east Asia among others, responded similarly to these powerful landscapes. Their places of worship were modelled on the most striking features of geology in which temples, churches, monasteries and cathedrals were orientated to the mountains or to pillars of stone, or shaped to resemble them with their interiors re-creating a 'cave-like environment' and with water used in the rituals performed inside (Taçon 1999). Like other Mesoamerican peoples, the Maya were convinced that the Earth was sacred and animated; mountains were thought to be hollow, and caves were entrances to the 'Underworld'. Rich with symbolism, water springing from their inner parts was considered ritually pure. Even today mountains, caves and waters are the places where ceremonies, settlement and social organization take place, while the centre of a Maya village is always marked by a cave (Brady and Ashmore 1999).

For prehistoric North Americans the most important rituals and ceremonies were related to:

(a) an 'ancestor cult' which refers to rituals that provide continued access to the deceased in the 'afterworld'. The ancestors were, in this way, engaged in the social, political and economic relationships of their descendants. Ahern (1973) and Watson (1988) affirm that they were used to both legitimize relationships among the living and the property of the land; and
(b) 'a fertility cult' concentrated in peripheral positions and readily transformed into pilgrimage centres where water was always present (Emerson 1977a; 1977b).

The barbagia di seulo landscape

In the central part of Sardinia Island (Fig. 1), a systematic survey for landscape features and for caves has produced good evidence for a Neolithic presence at several natural sites together with the hint that such special landscapes, in which primary and secondary burials were found, could have been considered 'sacred' by the people who inhabited them.

A sacred rock overhang—the 'Su Cannisoni'—and the 'Sa Omu 'e is ossus cave

South-west of the hill town of Seulo is a limestone cliff, 'Su Cannisoni'. Fig. 2 shows the rock overhang which is a naturally-formed horizontal recess in the limestone cliff (locally called *Tacchi* meaning *Heels* because of their flat shapes) from which a splendid view of Seulo and its surroundings can be admired.

The rock overhang is 20 metres from north to south, one to three metres wide and almost eight metres high. In the rock face are several prominent features in the form of vertical natural grooves or clefts. One is pink in colour and several metres long (Fig. 2). The others are smaller and located in the more western edge of the rock shelter. All these clefts resemble feminine physiology in the form of a huge 'yoni'. Yoni is Sanskrit/Hindi for vulva where in India the goddess Kunti is still worshipped. The site's local name is *Cannisoni*. This seems meaningless unless the word was anciently *Cunnisoni* in which case it may be a variant of the Latin 'kunnus' also meaning vulva . It is noticeable that in Sardo (the native Sardinian language) the word 'kunnu' is still used to refer, vulgarly, to a vulva. Hence, cunnus/cunnis could be the definitive word, either reaching Sardo via Latin or bypassing Latin in remote antiquity, such that here it indicates a sacred place (cunnus-yoni) where a fertility cult may have been ritually addressed to the Earth Mother.

The natural rock formation shown in Fig. 3 suggests, by using the concept of 'simulacrum', an image of a human head. It seems almost surreal because of its resemblance to the physiognomy of the local villagers.

Close to 'Su Cannesoni' is a cave named on old maps as 'Sa Omu è is ossus', which translates as 'The House of the Bones', but called in modern times 'Sa grutta 'e is bittuleris'. Because the cave served as an occasional shepherd's shelter, a low dry-stone wall had been built at its entrance (Fig. 4).

The deep dusty floor was filled with broken pottery and bones of human and animal origin (Fig. 5). Unfortunately within two years of its discovery by G. Gradoli in 2005 most of the bones were 'removed' by villagers unknown. In the lateral parts of this shrine, additional human bones and cranial parts were buried. There were more in natural holes in the rock of the upper part of the limestone hill, just above the rock overhang, which was full of bones disposed in a secondary burial with ceramic sherds (Fig. 6).

A rock tunnel in the river

The next site in this inquiry into the features of the Seulo

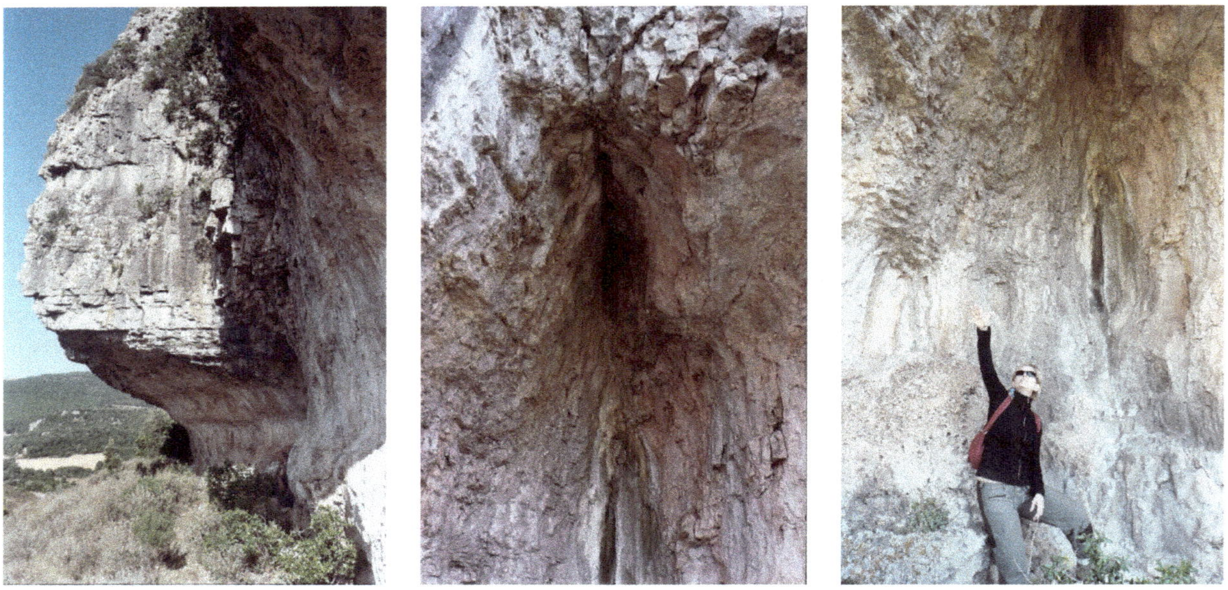

Fig. 2. The west-facing rock shelter and the spectacle of the pinkish cleft feature sheltered by the overhang.

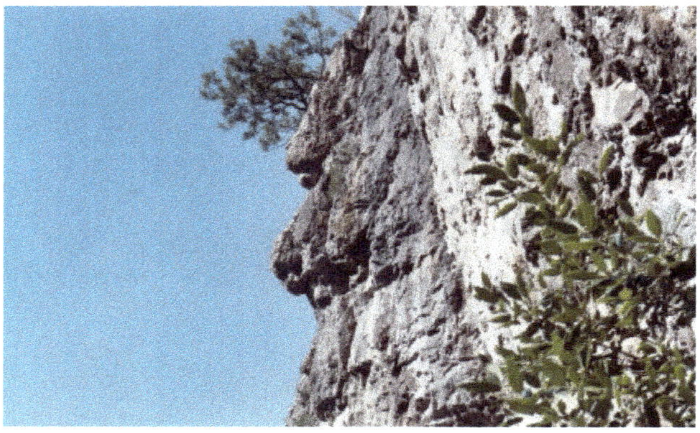

Fig. 3. Two views of the natural image of a head above the level of the rock overhang and 20 metres to its south.

ritual landscape is a rock tunnel 'Is Caddaias' through which passes the fast flowing water of the River Flumendosa. In the Sardu language the place-name means 'The Pools'. The site comprises three natural circular pools, located at different levels with waterfalls linking one to the next. The second pool has a 'rock tunnel feature' through which one can swim underwater to reach the last one. The place is well known by people of central Sardinia and, before its partial destruction by excessive floodwater in the winter of 2004/2005, local people used to bathe and show their bravery by swimming several times through it (Fig. 7a, b, c). The manner in which this underwater feature was used hints at the tradition being a former ritual and long-time custom linked to Earth Mother rustic beliefs. The suggestion follows the explanation known for an initiation/rebirth concept from the East Mojave Desert in California (and other sites). The hole in the Amer-Indian rock is a passage (dry in this case) that was used so much by Great Plains Indians that its internal floor was rendered smooth (Krupp 1997, 107-108).

'Su Stampu 'e su Turrunu': Another rock shelter with female features

The fourth site is a rockshelter with cascade and pool (Fig. 8) locally called 'Su Stampu 'e su Turrunu' or 'Su Stampu 'e su Turnu'.

It is situated at the municipality boundary between the villages of Sadali and Seulo in the inner part of the Forest of Addolì. The photographs of Fig. 8 were taken in the summer dry season. In winter (Fig. 9) the waters of a stream fall into a first pool that is not visible from the lower part where the photograph (red no. 2) was taken, pass into an opening (marked with a red 3) and go down through a yoni-like rock tunnel (marked with a red 2) where they collect at the base of the rock shelter (number 1) forming a huge water pool (Fig. 9).

The name of the site in Sardo means *'The hole of Turnu* or *Turrunu'* but the noun *'Turnu* or *Turrunu'* has no recognisable meaning, so would have been inherited from a previous ancient unwritten language different from the Sardinian language. Among remote possibilities is a deity figure related to the widely known Tara or Taran.

The site Su Stampu 'e Su Turrunu is well known all around the Barbagia di Seulo in central Sardinia. Local shepherds and inhabitants have always viewed it as a 'special place'. For this reason, for several decades and even now, both Municipalities of Seulo and Sadali claim the property of the site. Close by, fragments of obsidian—likely tool-making debitage—were located by the authors. The diameter and

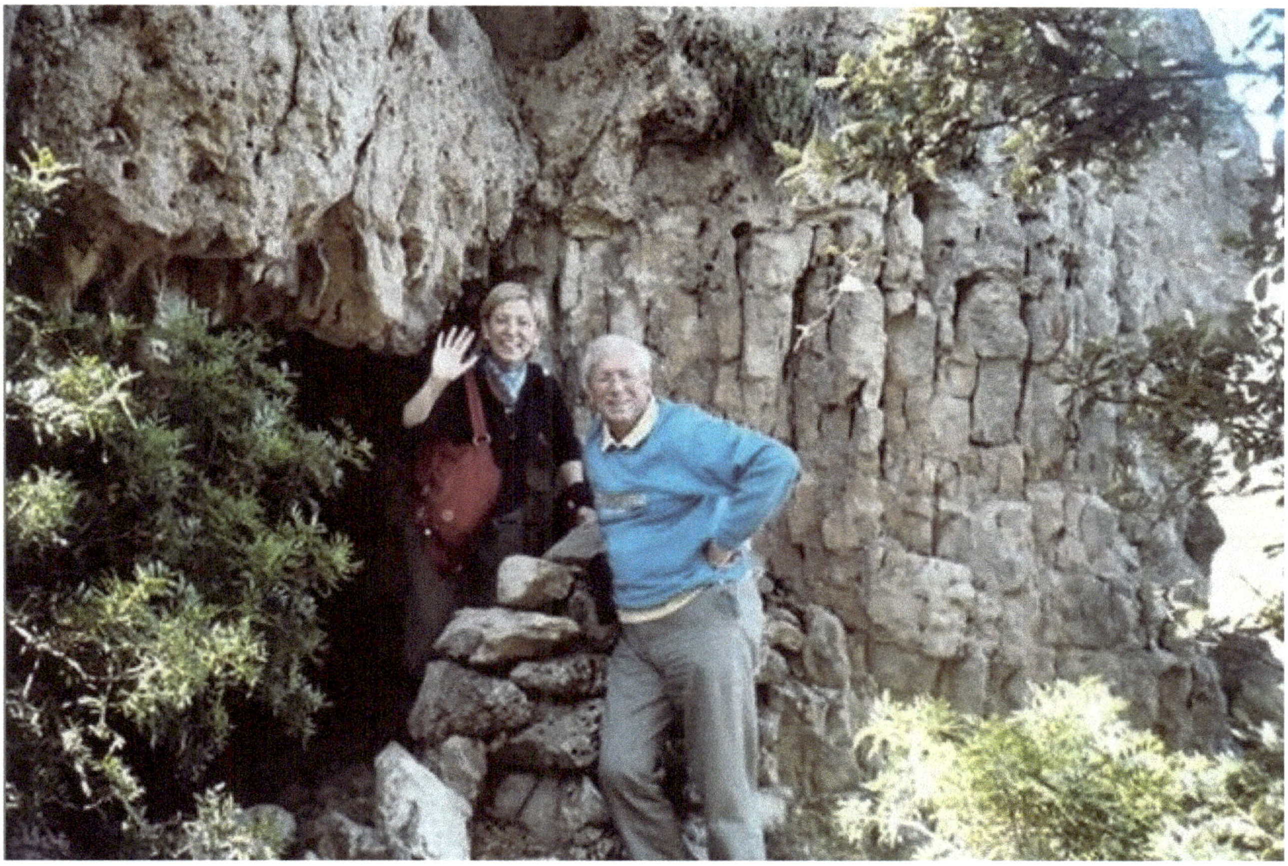

Fig. 4. The entrance to Sa Omu 'e Is Ossus with its shepherd's stone wall.

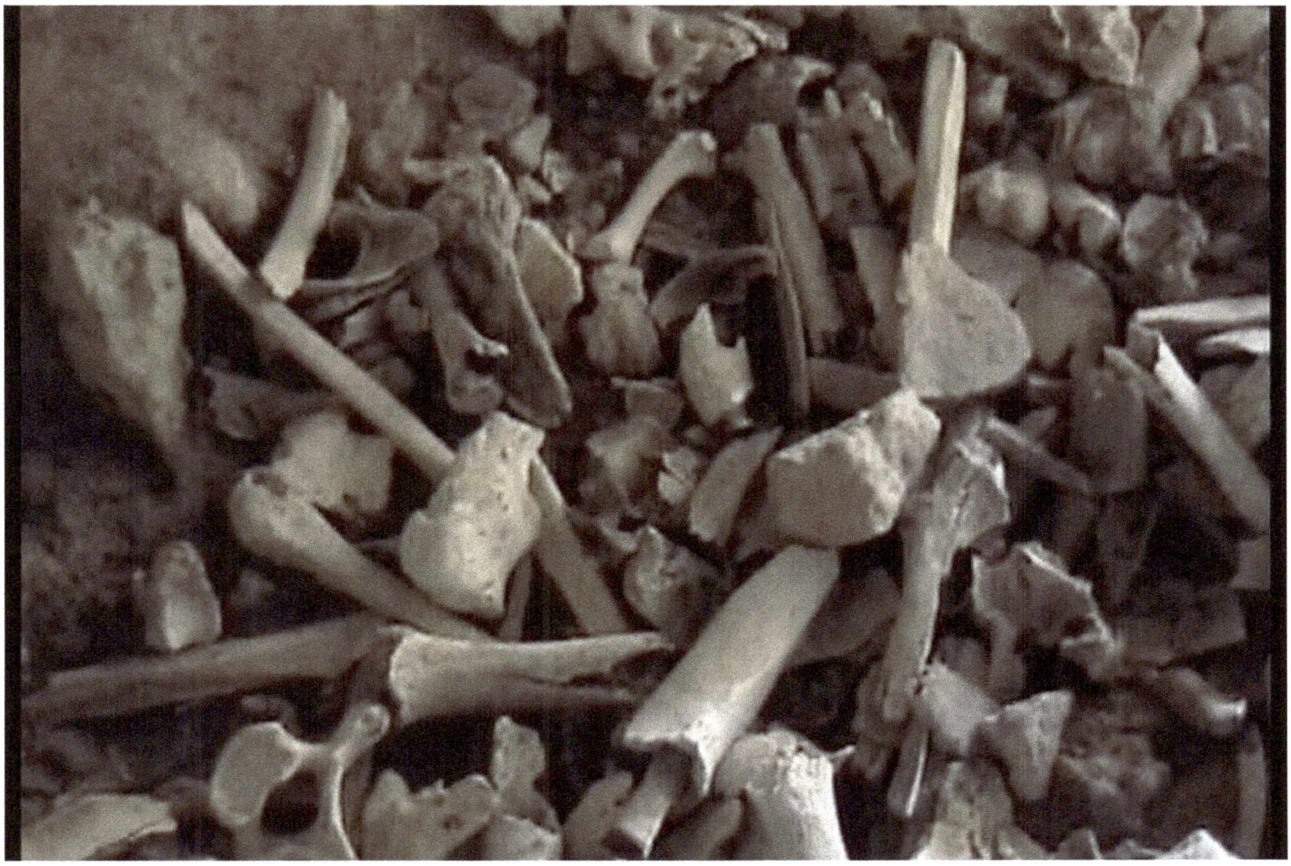

Fig. 5. Part of the cave floor in Sa Omu 'e Is Ossus, as photographed by G. Gradoli in 2005.

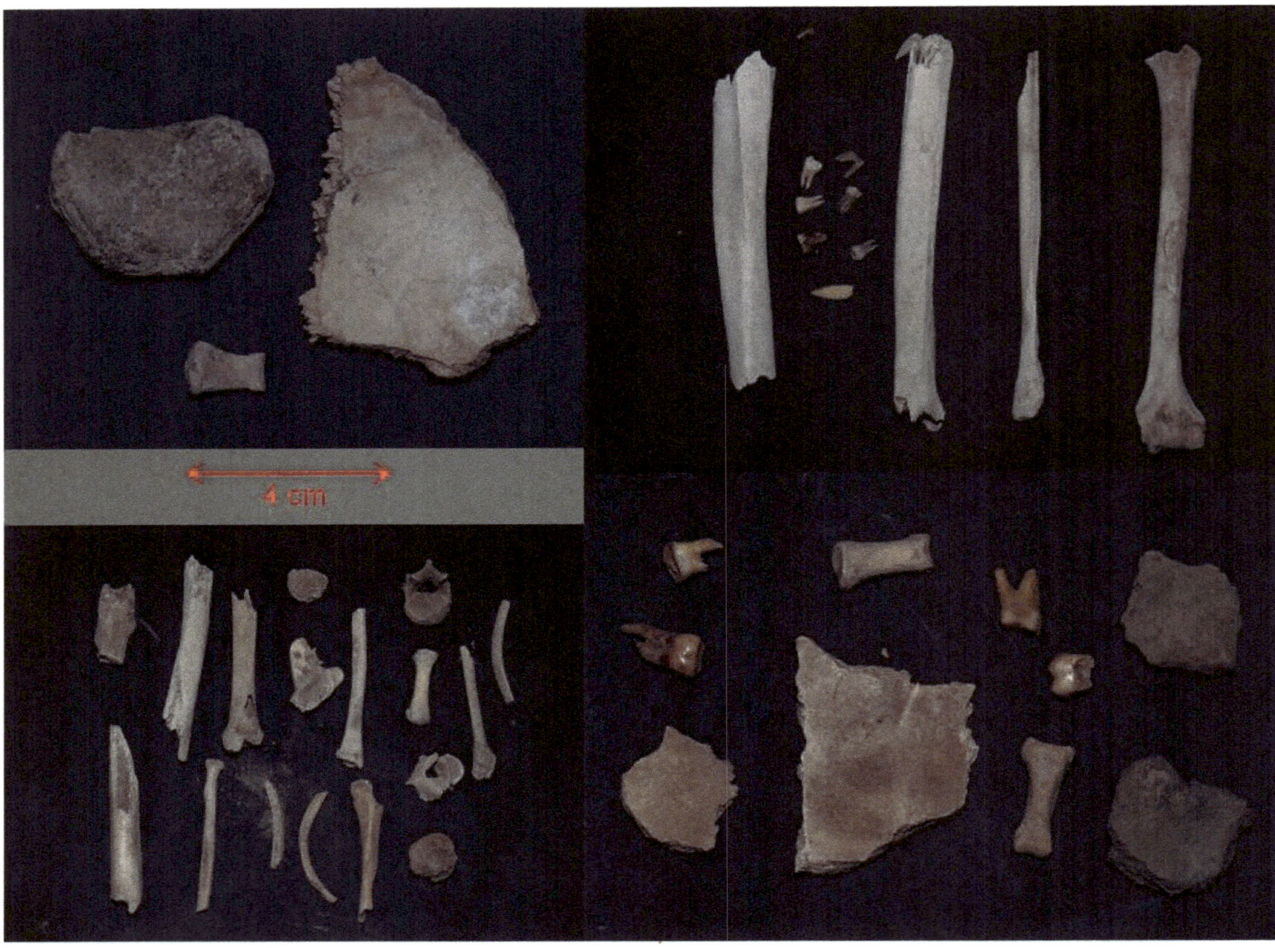

Fig. 6. Some of the human bones, animal bones and pottery from the cave Sa Omu 'e Is Ossus

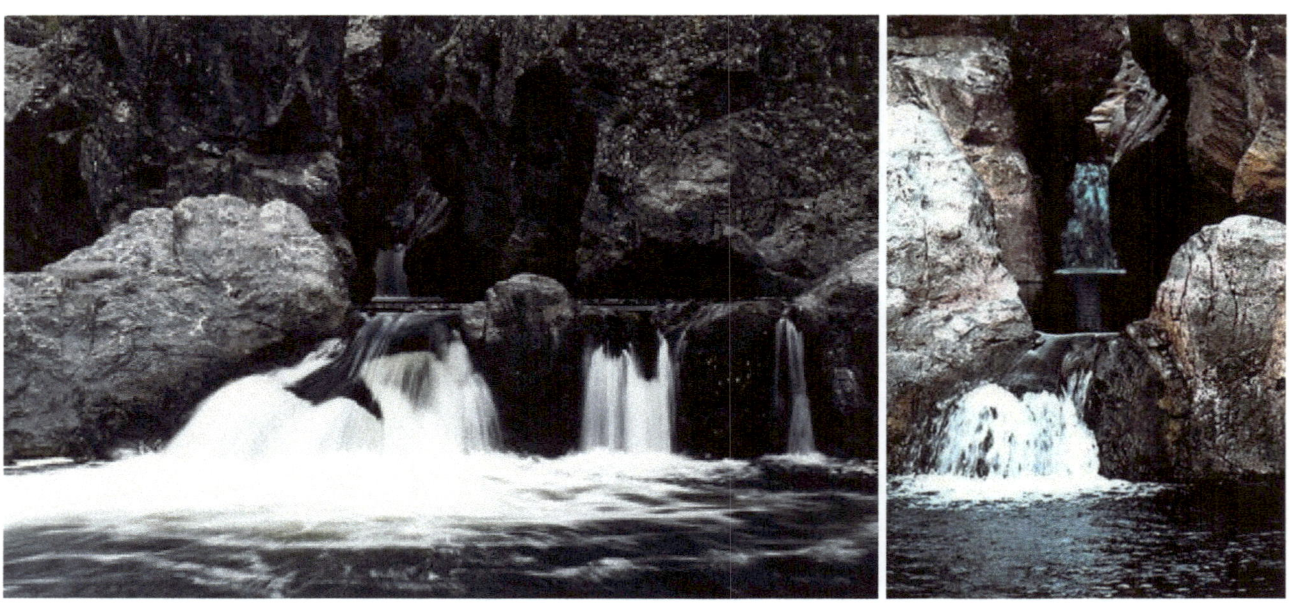

Fig. 7. The Flumendosa with the rock tunnel Is Caddaias.

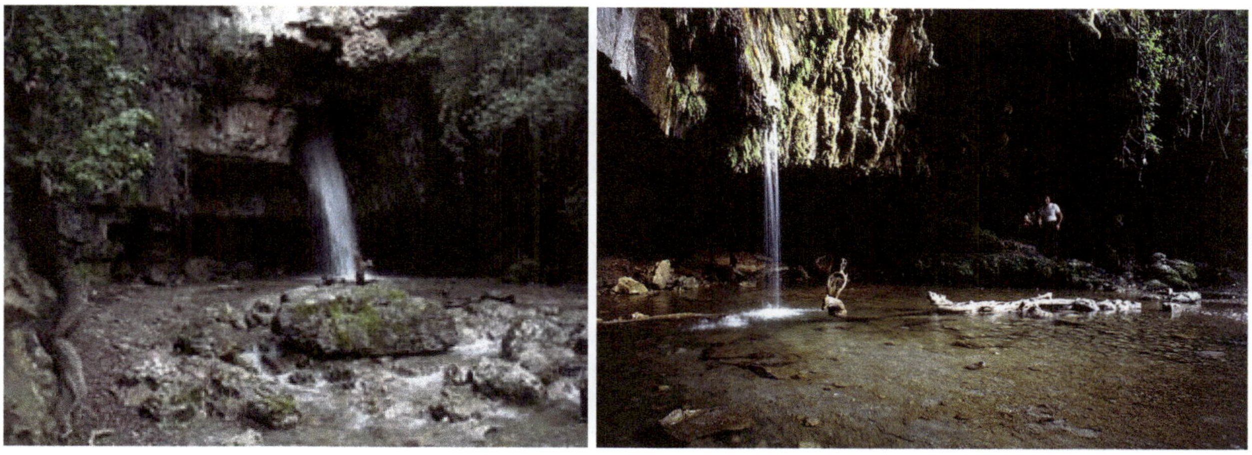

Fig. 8. Su Stampu 'e Su Turrunu.

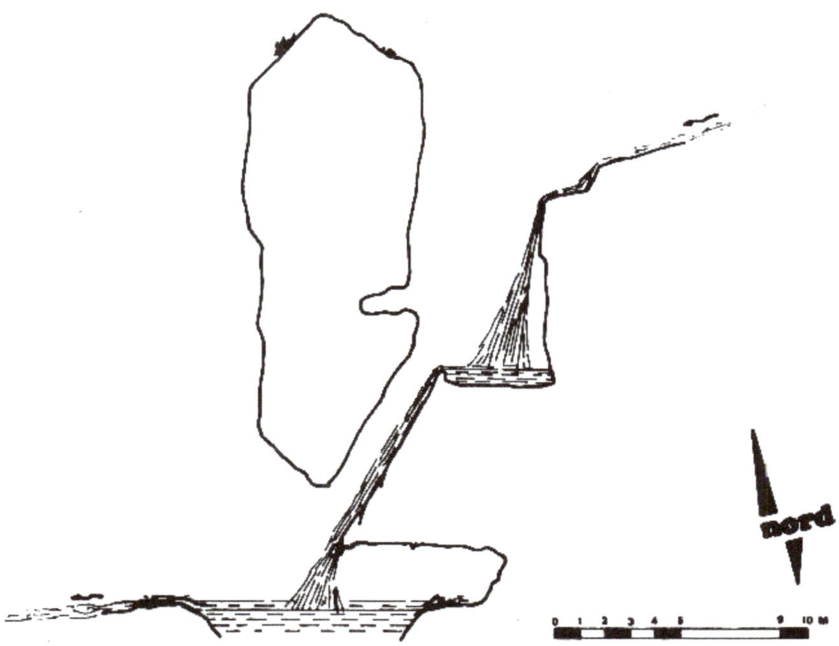

Fig. 9. Su Stampu 'e Su Turrunu (schematic side view).

depth of the pool are seasonal. In summer the lower water level allows two stakes to appear that had been driven into the rocky pool a long time ago (Fig. 10). The stake that was dated at Oxford University gave a radio-carbon age in the 18th or 19th centuries AD (Fig.11).

The second radio-carbon dating estimates the age of a tree-branch that has lain in the pool for many years (Fig. 11 and 12).

Concluding remarks

A previously unrecognised Neolithic landscape has been discovered in the mountainous region of Barbagia di Seulo in Sardinia. Features include rock shelters, caves, fissures, and an underwater rock tunnel—together with considerable

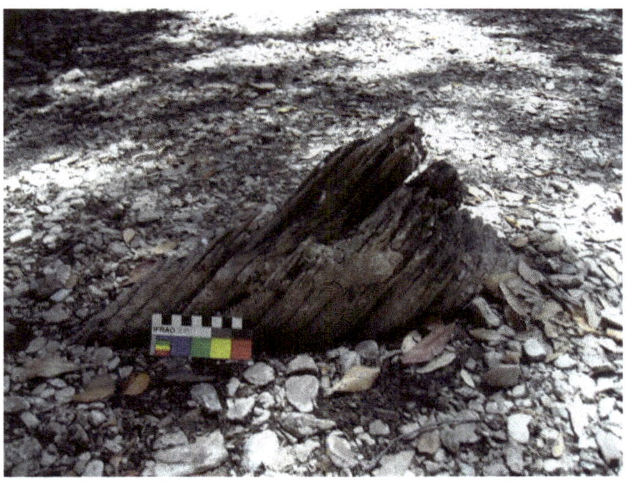

Fig. 10. The remains of one of two stakes

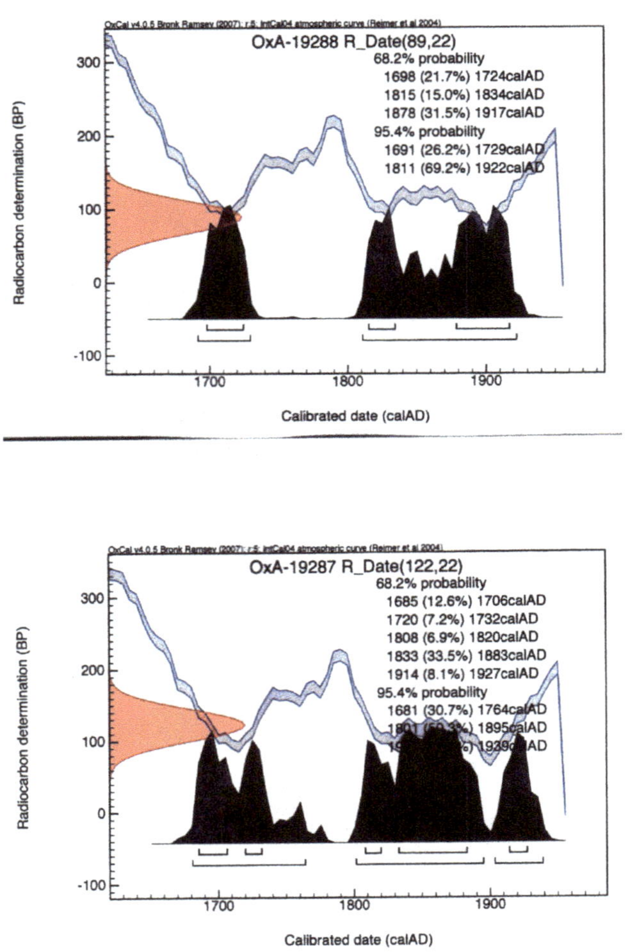

Fig. 11 (lower) Radiocarbon dating of a wooden stake. (Upper) Radiocarbon dating of a sample from the tree branch.

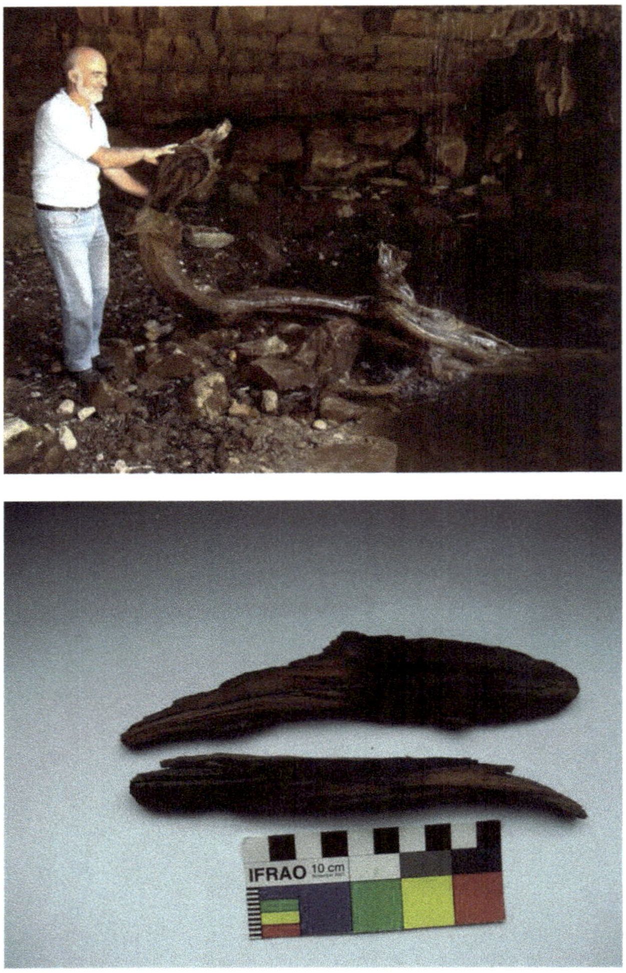

Fig. 12. The old tree branch and fragments used for dating.

quantities of animal bone, human bone, potsherds, and some tools. As with other cultures studied across the world one infers that the territory was deemed to be sacred or it had potential for sanctity at certain times of the year depending on the nature of the human activity and supposed presence of the divine. Likewise, the territory probably served as a cultural mentor (Basso 1984, 1996), and at selected natural places shrines were set up in order to focus and strengthen memories and to reinvigorate appropriate activities. More detailed information on two caves follows in the next two papers.

References

AHERN, M. (1973). *The cult of the Dead in a Chinese Village.* Standford: Standford University Press.

ALDENDERFER, M. (1990). Late pre-ceramic ceremonial architecture at Asana, Southern Peru. *Antiquity,* 64, 497-493.

ASHMORE, W. and KNAPP A. B. (eds.), 1999. *Archaeologies of Landscapes. Contemporary Perspectives.* Blackwells, Oxford.

BASSO, K. H.(1984). Stalking with Stories: Names, places, and moral narratives among the Western Apache. In S. Plattner and E.M. Bruner (eds), *Text, Play, and Story: the construction and reconstruction of self and society.* 1983. Proceedings of the American Ethnological Society, 19-55. Washington, D.C.

BASSO, K. H. 1996. *Wisdom sites in place: Notes on Western Apache Landscape.* In S. Feld and K.H. Basso (eds), Senses of Place, Santa Fe, SAR Press, 53-90.

BRADY, J.E. and ASHMORE, W. (1999). Mountains, Caves, Water: Ideational Landscapes of the Ancient Maya. In Ashmore, W. and Knapp A. B. (eds), 1999. 124-140. *Archaeologies of Landscapes. Contemporary Perspectives.* Blackwells. Oxford.

BRADLEY, R. (1991). Rock Art and perception of the Landscape. *Cambridge Archaeological Journal,* 1, 77-101.

BRADLEY, R. (1997). *Rock Art and the Prehistory of Atlantic Europe.* London: Routledge.

BUIKSTRA, J..E. and CHARLES, D. K. (1999). Centering the Ancestors: Cemeteries, Mounds, and Sacred Landscapes of the Ancient North American Midcontinent. In ASHMORE, W. and KNAPP A. B. (eds), 1999. *Archaeologies of Landscapes. Contemporary Perspectives.* 201-222. Blackwells, Oxford.

CARMICHAEL, D.L., HUBERT, J., REEVES, B. and

SCHANCHE, A. (eds) 1994. *Sacred Sites, Sacred Places.* Routledge. London

CRUMLEY, C.L. (1999). Sacred Landscapes: Constructed and Conseptualized. In Ashmore, W. and Knapp A. B. (eds), 1999. 269-275. *Archaeologies of Landscapes. Contemporary Perspectives*. Blackwells. Oxford.

EMERSON, T.E. (1977a).*Cahokia and the Archaeology of Power*. Tuscaloosa: University of AlabamaPress.

EMERSON, T.E. (1977b). Cahokian elite ideology and Mississipian cosmos. In T.R. Pauketat and T.E. Emerson (eds), *Cahokia: Domination and ideology in the Mississipian World.* Lincoln: University of Nebraska Press, 190-228.

GRADOLI, M.G. et al. (2005). Il binomio uomo-territorio nella Barbagia di Seulo (Sardegna centrale): prime segnalazioni di pitture parietali ed incisioni in grotte carsiche. Atti del XVI Convegno degli Antropologi Italiani. Genova. Edicolors.

GRADOLI. M.G. and MEADEN, G.T. (2012). "Discovery of prehistoric sites in caves and rockshelters in the Barbagia di Seulo, South-Central Sardinia", in K. A. Bergsvik and R. Skeates *Caves in Context: The economical, social and ritual importance of caves and rockshelters.* 188-197. Oxbow Books, Oxford.

GRADOLI, M.G. and MEADEN, G. T. (2012a). Neolithic art and artefacts in caves near Seulo in Central Sardinia: 'Grutta I De Longu Fresu'. In Meaden, G. T. (2012) *Archaeology of Mother Earth Sites and Sanctuaries through the Ages: Rethinking symbols and images, art and artefacts from history and prehistory*. British Archaeological Reports. Oxford.

GRADOLI, M.G. and MEADEN, G. T. (2012b). Prehistoric occupation and use of a cave, near Seulo in Central Sardinia: 'The Oval Room' at 'Is Janas'. In Meaden, G. T. (2012) *Archaeology of Mother Earth Sites and Sanctuaries through the Ages: Rethinking symbols and images, art and artefacts from history and prehistory.* British Archaeological Reports. Oxford

HAWKES, C. (1954). Archaeological theory and method: some suggestions from the Old World. *American Anthropologist*, 56, 155-168.

KRUPP, E.C. (1997). *Skywatchers, shamans and kings*. 107-108. Wiley, New York.

McEWAN, C. and VAN DE GUCHTE, M. 1992. Ancestral time and sacred space in Inca state ritual. In R. F. Townsen (ed.). *The Ancient Americas : Art from Sacred Landscapes*. 359-371. Chicago.

TACON, C. (1999). Identifying Ancient Sacred Landscape in Australia: from Physical to Social. In Ashmore, W and Knapp A.B. (eds), 1999. *Archaeologies of Landscapes. Contemporary Perspectives*. 33-53. Blackwells. Oxford.

UCKO, P.J. (1994). Foreword. In D. L. Carmichael, J. Hubert, B. Reeves, and A. Schanche (editors). *Sacred Sites, Sacred Places*. One World Archaeology, 23, London: Routledge, xiii-xxiii.

WATSON, R. S. (1988). Remembering the dead: graves and politics in south-east China. In J.L. Warson and E. Rawski (eds), *Death, Ritual in Late Imperial and Modern China*. 203-227. University of California Press. Berkeley.

Neolithic Art and Artefacts in Caves Near Seulo in Central Sardinia: 'Grutta I De Longu Fresu'

M. Giuseppina Gradoli

COMET – Valorizzazione Risorse Territoriali and ISSEP Sardinia, Via dei Colombi, 31, 09126 Cagliari

G. Terence Meaden

Oxford University, Department of Continuing Education (Archaeology) and Kellogg College, 62 Banbury Road, Oxford, England, OX2 6PN

***Abstract**: Two caves in the Barbagia di Seulo region of mountainous Central Sardinia were identified by their deposits as having been used in the Middle Neolithic Age. One is the 'Grutta I De Longu Fresu', sited near a river whose name translates as 'The Trees of the Dead'. A panel of rock art, thickly overlain with stalagmite flow, was found. The black paintwork appears to depict anthropomorphic figures that may relate to traditional folklore activities that still take place in villages of central Sardinia today. A human skull at the far end of this cave's main gallery is also sealed by stone-flow, and was dated at Oxford University to 4259-4042 cal. BC (at 95.1% probability, OxA-X-2236-44). This is the oldest-dated Neolithic human remains known for central and southern Sardinia. The other cave (the Oval Room at Is Janas) is treated in a separate paper.*

***Key words**: Neolithic, rock art, Sardinia, symbolism*

***Riassunto**: Nella regione montuosa della Barbagia di Seulo, Sardegna centrale, sono state identificate due grotte con depositi archeologici ed antropologici del Neolitico Medio. Una è la Grotta 1 de Su Longu Fresu, situata in prossimità di un ruscello il cui nome indica, nel dialetto locale, 'l'albero dei morti'. Una parete con pitture rupestri di tipo antropomorfo e zoomorfo, di colore nero, sigillate da un velo di carbonato di calico è stata rinvenuta all'estremità destra della cavità. Nella parte finale della grotta un cranio umano, ancorato al suolo da uno spesso deposito carbonatico, è stato datato nei Laboratori dell'Università di Oxford avente 4259-4042 anni prima di Cristo (probabilità del 95.1% , OxA-X-2236-44). La seconda grotta sarà trattata separatamente in un altro lavoro.*

***Paroli chiave**: Neolitico, Sardegna, pitture parietali, simbolismo.*

Introduction

This is the second of three summaries about discoveries made in the mountainous region of the Barbagia di Seulo in central-southern Sardinia. A study of place-names and a recognition of landscape features led to the discovery and interpretation of a number of Neolithic sites of caves and rock shelters (Gradoli 2005; Gradoli and Meaden 2012, 2012a, 2012b). Similarity of certain symbols may have developed worldwide because of fundamentals in the psyche that are common to agricultural communities everywhere through respect for ancestors and the people's need for fertility success in their fields, their animals and their women.

The cave described in the present paper is the 'Grutta 1 De Longu Fresu' with its painted rock art. A well-referenced archaeological summary of Sardinia's caves is given by Skeates (2012). This is one of the first examples of prehistoric painted rock art known for all Sardinia, although mural rock art painted as spirals on subterranean walls has since been reported (but not yet published) from a site 100 km to the west near the coast.

The cave 'Grutta I de Longu Fresu'—and its art and shrines

The cave 'Grutta 1 De Longu Fresu', which we call *"Skull Cave" following our major discovery*, opens in a direction NNE—SSW above a small stream called *'Riu Su Longu Fresu'*. In the Sardinian language, Sardo, the *'Longu Fresu'* is a species of tree that is said, locally, to be the 'plant of the dead'. The cave is naturally formed in the limestone cliff. It is about 15 metres long, 2 to 3 metres wide and 1 to 2 metres high. Fig. 1 is a plan. Fig 2 is a general view.

At the far end, opposite the entrance, a human half-skull anchored in stalagmite (Fig. 3) was positioned near a natural hole in the end-wall which, when cleared of the

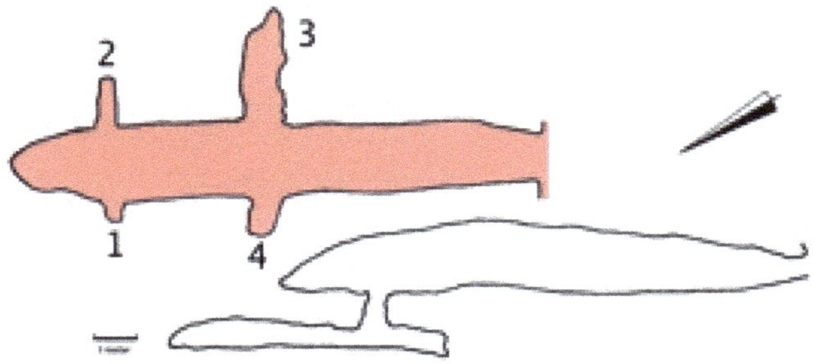

Fig. 1. Plan of the 'Grutta I De Longu Fresu' cave (pink) with its niches numbered 1 to 4, and a side view (white) drawn as if looking from the north-west.

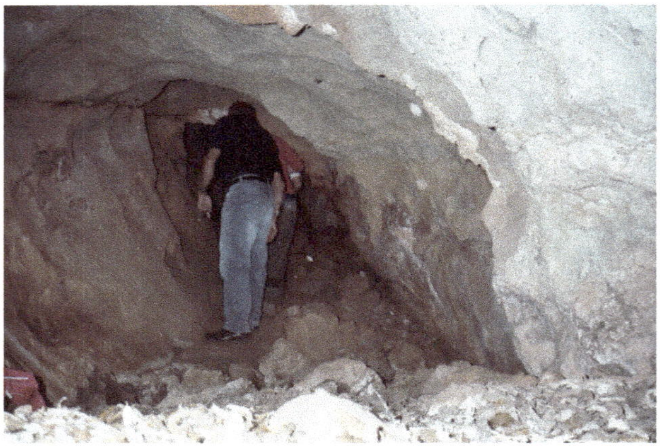

Fig. 2. View SSW along main passage.

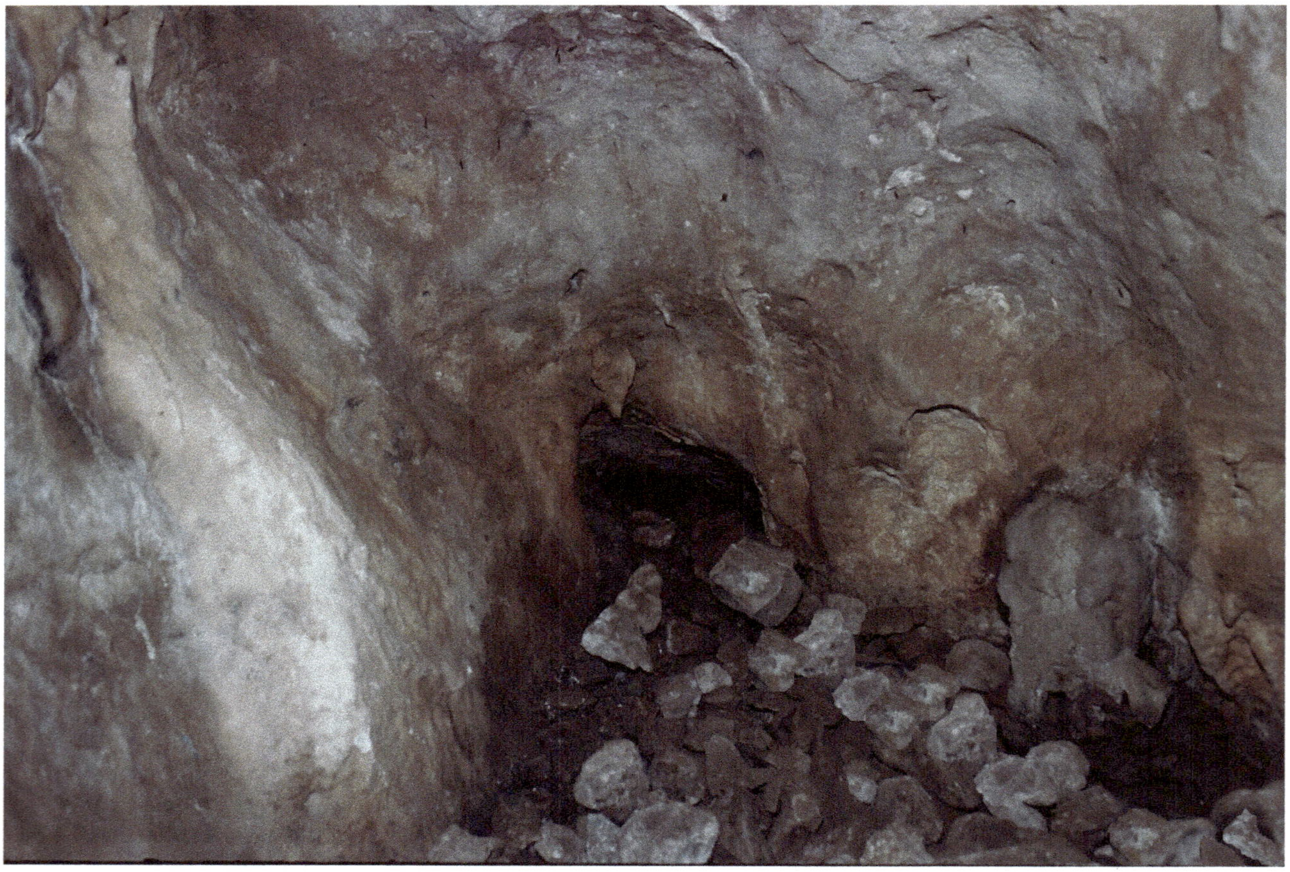

Fig. 3. Human skull (at right) and the hole in the cave wall that held the skull pieces of Fig.4.

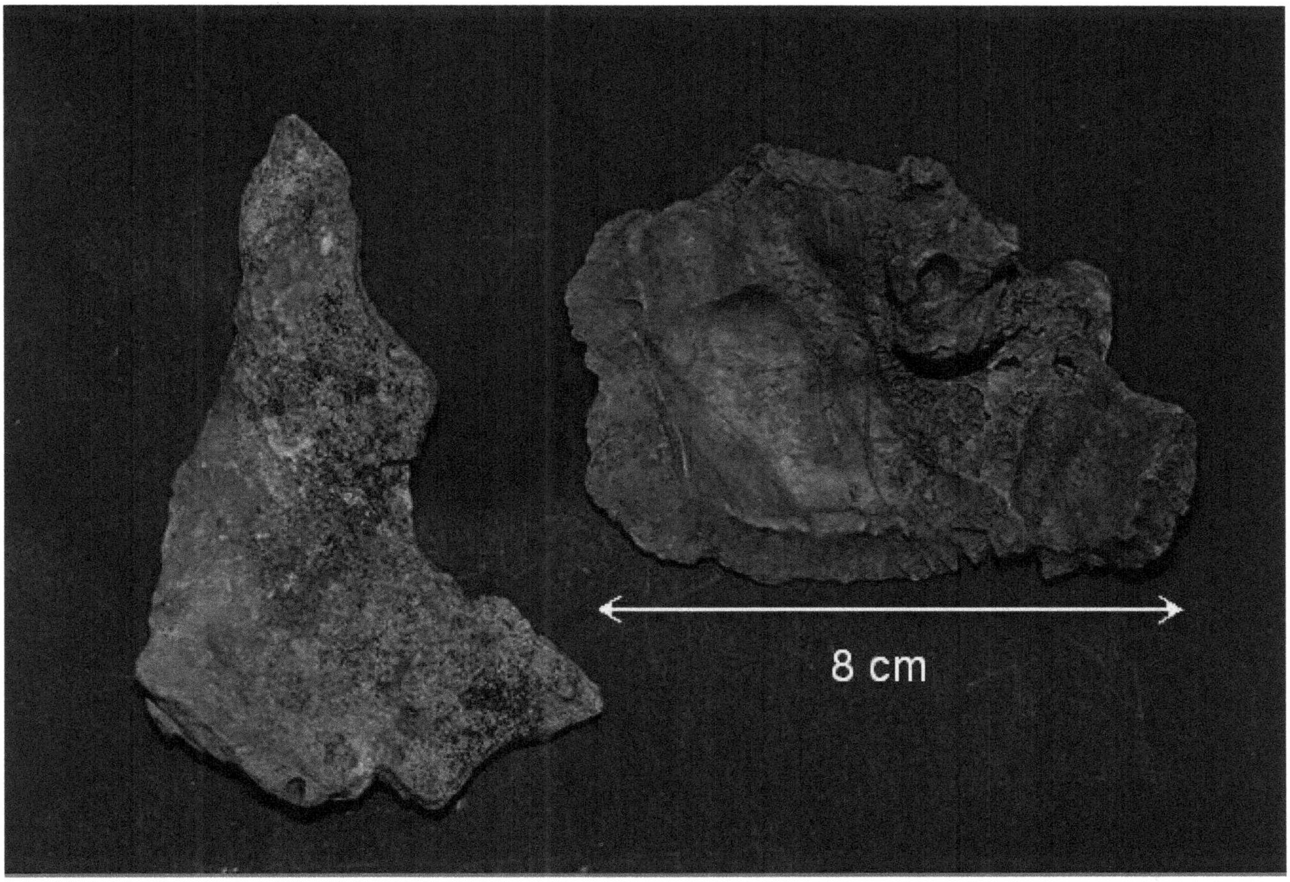

Fig. 4. Pieces of skull from the hole pictured in Fig. 3

little stones that had closed it, revealed additional pieces of a cranium (Fig. 4).

Fig. 5 shows a lateral view of the skull. This was sampled and analysed at the University of Oxford Radiocarbon Unit, and was dated to 4259–4042 BC (95.1% probability, OxA – X – 2236 – 44) (Fig. 5). This is Middle Neolithic and the oldest human skull known for central and southern Sardinia.

The stone-covered cave floor has since been excavated to determine whether artefacts and more human bones are buried beneath it. On the plan of the cave (Fig. 1), the numbers 1 to 4 indicate natural alcoves or niches (e.g. Figs. 6, 7) created long ago by the passage of falling waters. Niches 2 and 4 produced more discoveries of Neolithic Age.

Niche number 2

Niche 2 presents at its base a 70-cm deep artificially-excavated pit (Fig. 7) which could have served to collect water draining into a lower level of the cave in order to prevent flooding of the skull and other bones. On the left side of this niche, a 30cm by 30cm rock-art panel was drawn by the finger of an adult using natural black paint (Fig. 8) which was preserved by a thin layer of calcium carbonate. The latter will be dated by sampling at the base of the stalagmite that coats it using the Uranium Isotopic Series.

Use of the colour black is apposite—because it befits an underworld cave-shrine to death and the ancestors. Black symbolizes death, whereas the colour more often encountered in rock art in external different circumstances is red ochre, which symbolizes life. For instance, compare Simek and Holliday (2009) with regard to the Amer-Indians of the South-West Plains.

In Fig. 9 the photograph of the black pictographs have been deliberately *"colour enhanced"*. Each different colour represents a continuous movement of the finger in the act of drawing. In green one can see an anthropomorphic horned-head figure (possibly wearing a wooden carved mask typical of the Sardinian pastoral tradition) and holding a bow and arrow towards a horned animal depicted in red. The others signs, in part discoloured and incomplete, have not been interpreted.

Niche number 4

In niche number 4 (Fig. 10) pieces of human skull were found beneath stones covering the floor. Here, too, are features of ancient waterfalls and a pit excavated at its base.

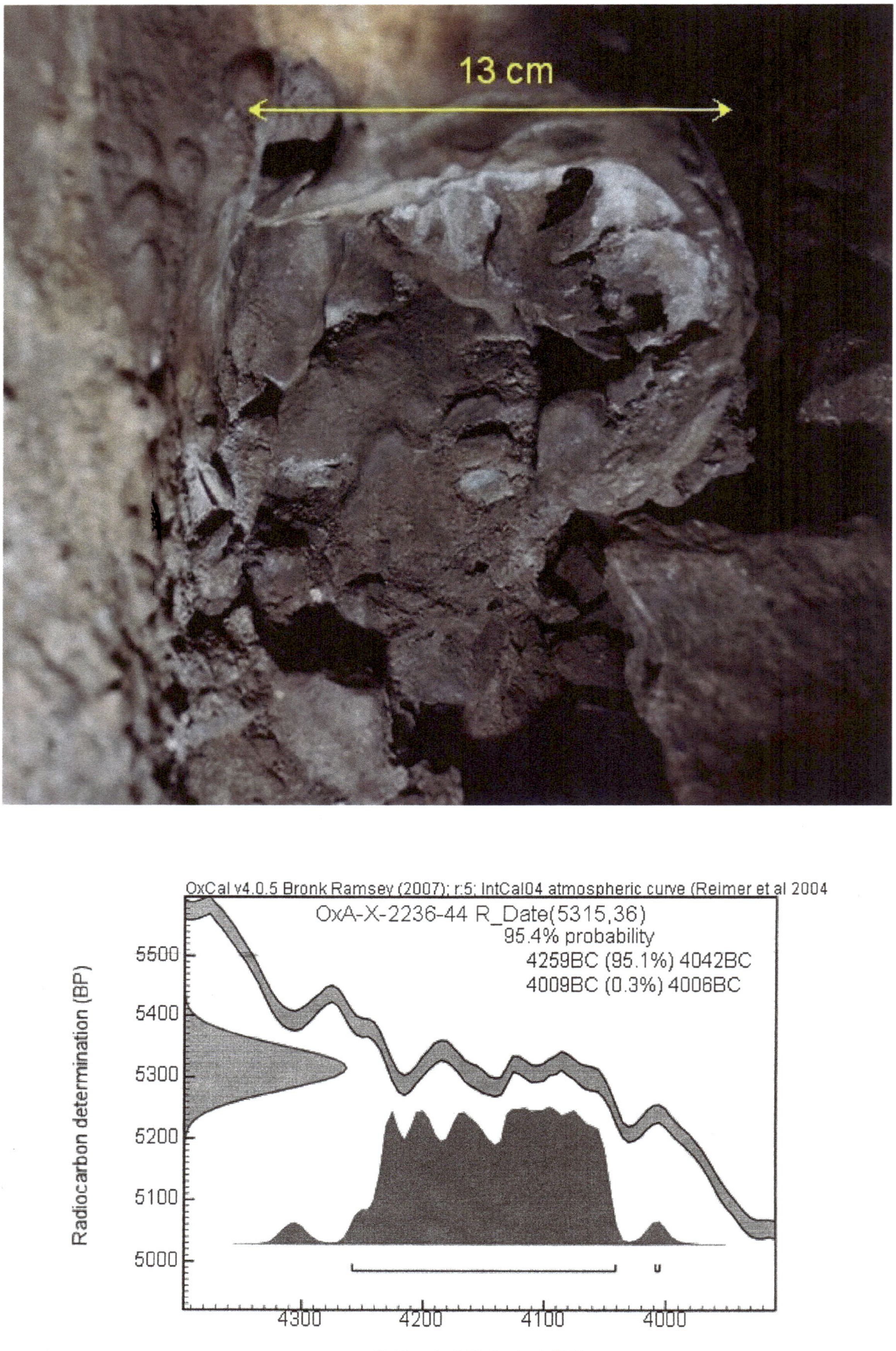

Fig. 5. Close-up of the stalagmite-covered skull dated to about 4259-4042 BC

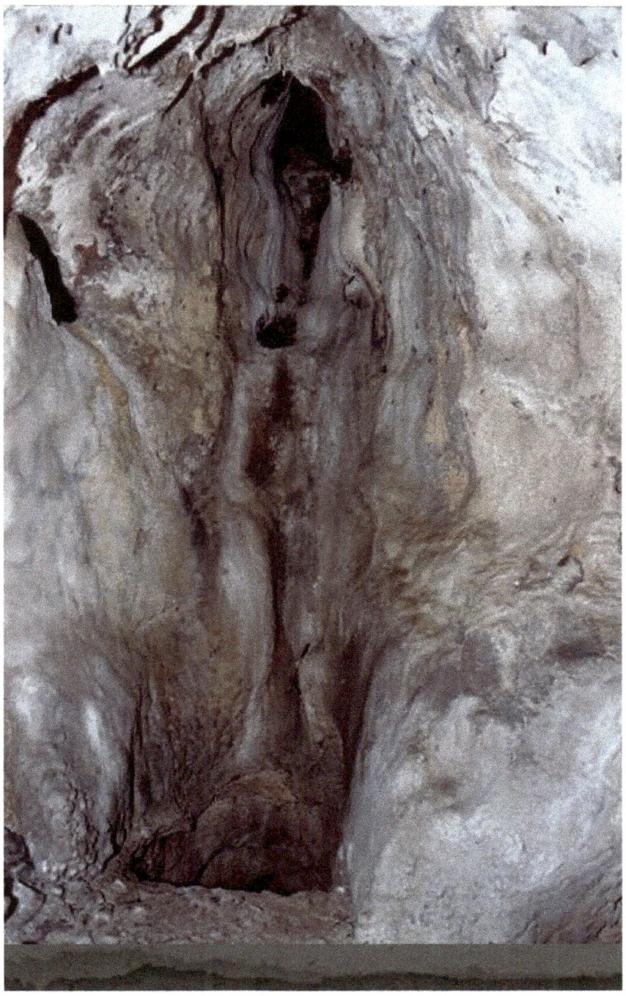

Fig. 6. Alcove or niche number 2. Note the black-painted marks on the cave wall at the left

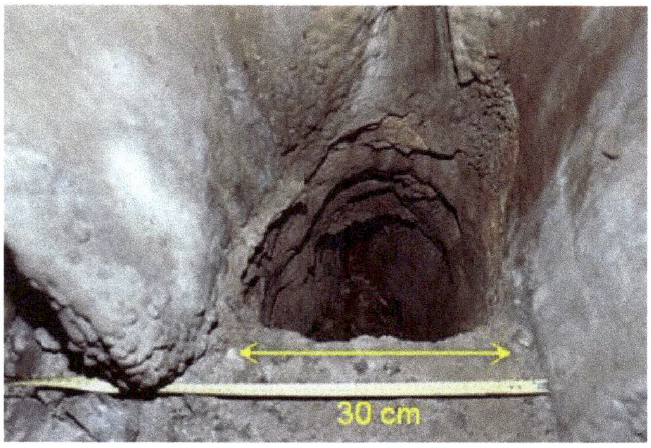

Fig. 7 Vertical picture illustrating the artificially-deepened hole at its base.

Comparing 'Grutta I de Longu Fresu' with 'West Kennet Long Barrow' in England

Some previous researchers have commented that chambered long barrows are artificial constructions whose primary objective was to imitate the dark interior of natural caves. Cyriax (1921) made a case that the interior of passage barrows and chambered barrows (i.e. artificial caves) represented the dark womb of a mother deity like the Earth Mother.

The long barrow at West Kennet near Avebury, Wiltshire, England, was excavated by Piggott (1962). Fig. 11 is from his plan. In the previous century Dr. Thurnam (1861, 416) entered the barrow by digging through the roof of the western end. One of several discoveries was the skull of a child positioned on the floor at the western end against the back wall on the centre line of the gallery (Meaden 1999, 102-108). This is wholly similar to the placing of the human skull at the far end of the cave of Grutta I de Longu Fresu. Might there be a similarity of purpose for these two ancestral/death locations that are so far apart in Europe and yet so close in Early-Neolithic agrarian/pastoral prehistory? The ancient British deposited the skull in the barrow in about the 38th-century BC. (Whittle et al. 2008). The ancient Sardinians did so in the 42nd-century BC.

Conclusions

People appear to have made pilgrimages to the cave of Grutta I de Longu Fresu to honour the dead. They decorated a wall in a way that was meaningful to them but can only be guessed by us. There is a possibility that the art includes a human figure in ceremonial or magical headdress—and, indeed, current traditional events in the villages may be reminiscent of and relate to this. The placing of a human skull at the remotest spot in the cave—subsequently preserved beneath calcitic stoneflow—is similar in location to that known for a major Early-Neolithic British long barrow. Whatever else, the 20m-long natural cave named 'Grutta I de Longu Fresu' certainly became a shrine in Early Neolithic times over 6000 years ago. Excavations, led by Dr. Robin Skeates (2009-2010), began in the 2009 summer season.

References

BERGSVIK, K.A. and SKEATES, R. (2010 or 2011). *Caves in Context: The economical, social and ritual importance of caves and rockshelters.* Oxbow Books, Oxford.

CYRIAX, Tony (1921). Ancient burial places. *Antiq. J.* vol. 28, 205-215.

GRADOLI M.G. et al. (2005). Il binomio uomo-territorio nella Barbagia di Seulo (Sardegna centrale): prime segnalazioni di pitture parietali ed incisioni in grotte carsiche. Atti del XVI Convegno degli Antropologi Italiani. Genova. Edicolors.

GRADOLI, M.G. and MEADEN, G.T. (2011). 'Underworld and Neolithic Rituality: The rock art of the Su Longu Fresu Cave, Central Sardinia', Atti del XXIV Valcamonica Symposium nella sessione 'Art and Communication in Pre-Literate Societies', Capo di Ponte, 13-18 July 2011

GRADOLI. M.G. and MEADEN, G.T. (2012). "Discovery

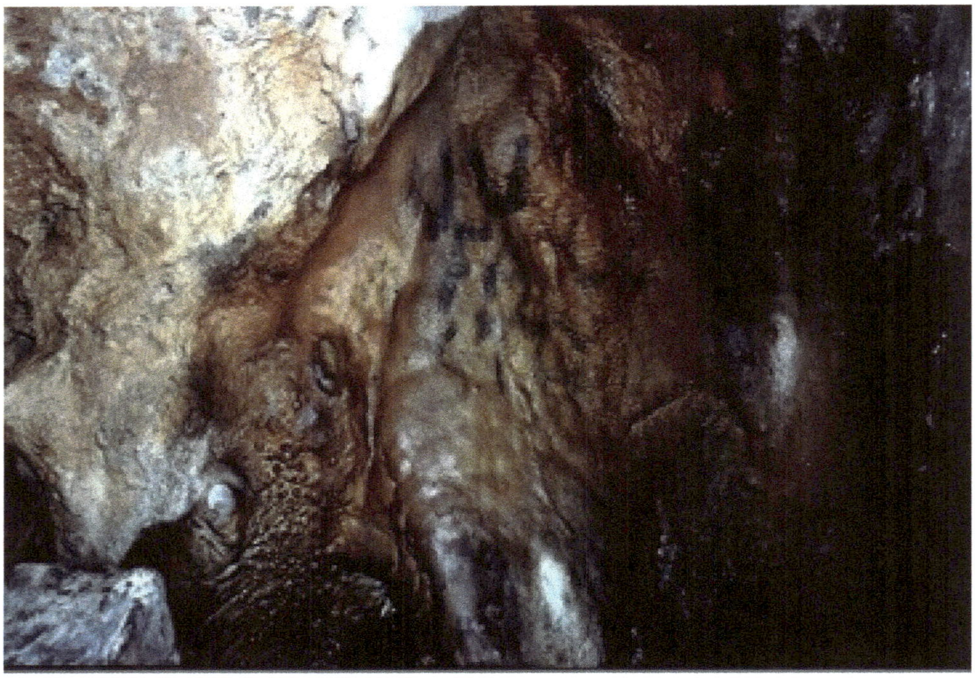

Fig. 8. The art: black drawings made by a finger

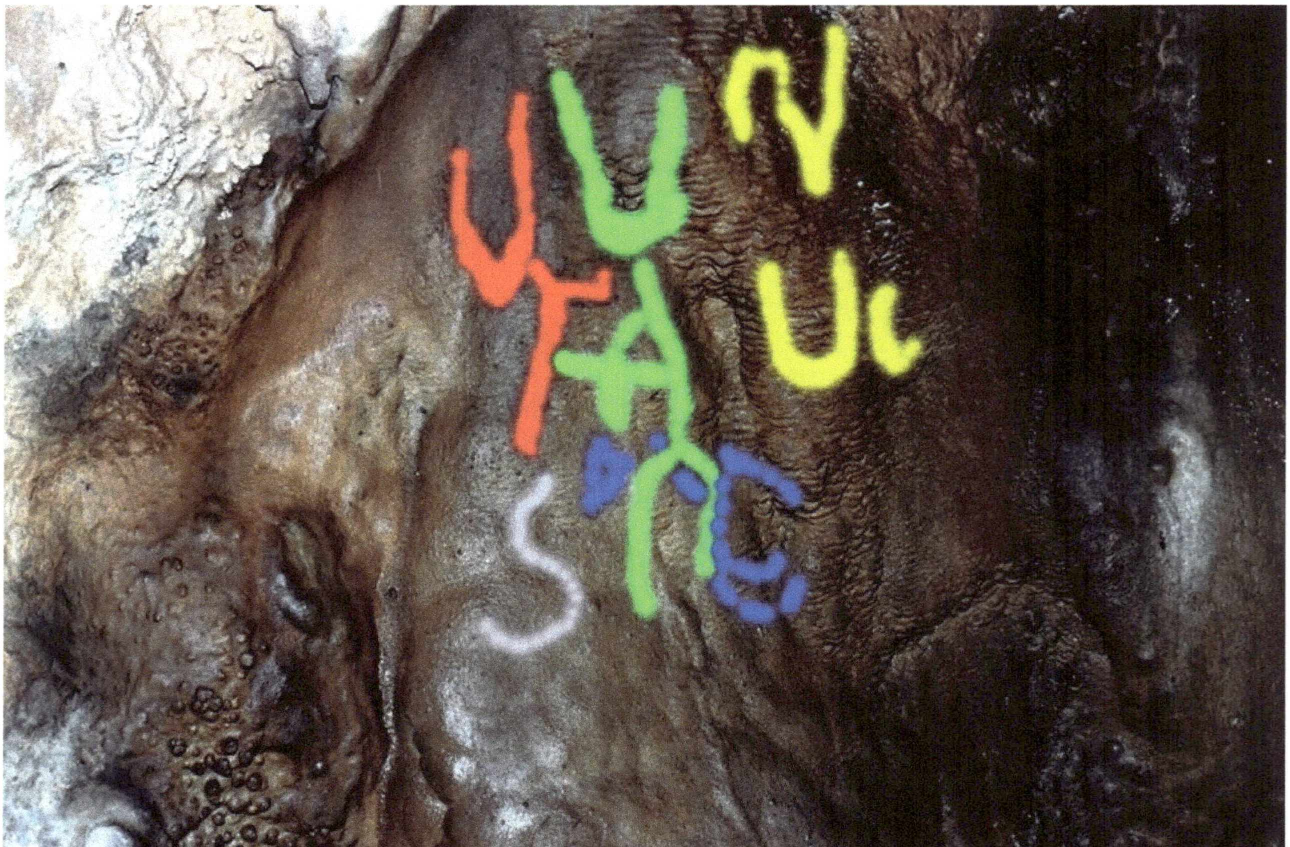

Fig. 9. Photograph of the same panel, highlighted in colour in order to emphasise the chief elements. The horned man holding a bow-and-arrow has been coloured green.

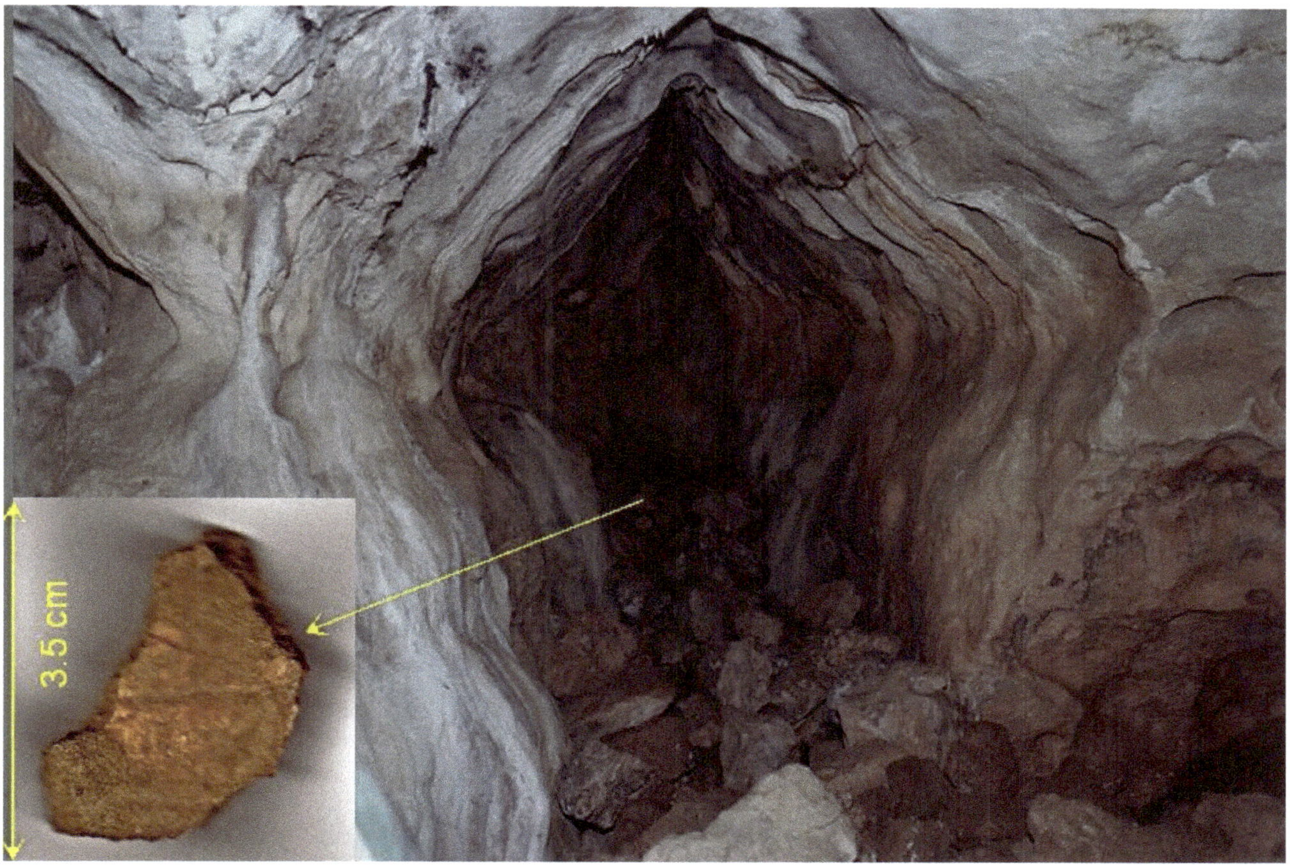

Fig. 10. Niche no. 4. In the foreground a piece of human skull found together with some pieces of ceramic in the inner part.

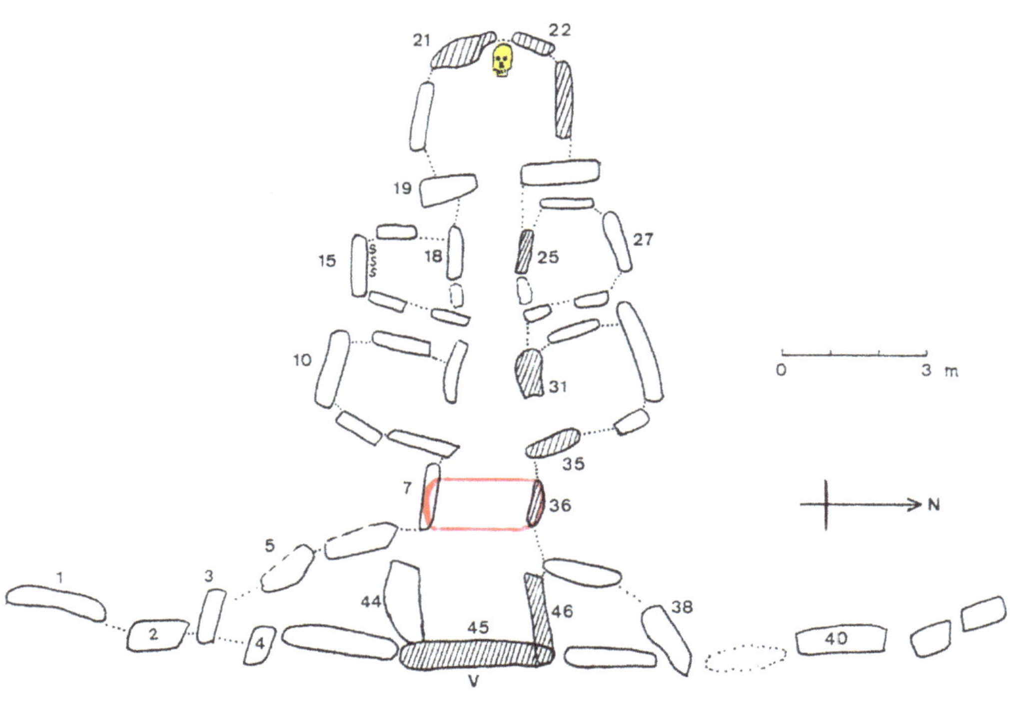

Fig. 11. Plan of the equinoctial long barrow at West Kennet, Avebury, England. The child's skull was positioned on the floor at the far western end. The letter V marks the stone carved with a vulva 2m long. The lintel between Stones 7 and 36 is carved as a deer.

of prehistoric sites in caves and rockshelters in the Barbagia di Seulo, South-Central Sardinia", in K. A. Bergsvik and R. Skeates *Caves in Context: The economical, social and ritual importance of caves and rockshelters.* 188-197. Oxbow Books, Oxford.

GRADOLI, M. G. and MEADEN, G. T. (2012a). Neolithic sacred sites in the mountains of Central Sardinia, In MEADEN, G.T. *Archaeology of Mother Earth Sites and Sanctuaries through the Ages: Rethinking symbols and images, art and artefacts from history and prehistory.* British Archaeological Reports. Oxford.

GRADOLI, M.G. and MEADEN, G.T. (2012b). Prehistoric occupation and use of a cave in the Neolithic landscape of Barbagia di Seulo, Central Sardinia: The Oval Room at 'Is Janas'. In MEADEN, G.T. (2012) *Archaeology of Mother Earth Sites and Sanctuaries through the Ages: Rethinking symbols and images, art and artefacts from history and prehistory.* British Archaeological Reports. Oxford.

MEADEN, G. T. (1999). *Secrets of the Avebury stones,* 102-108. Souvenir Press. London.

PIGGOTT, S. (1962). *West Kennet Long Barrow: Excavations 1955-56.* Plans pp. 10, 22. H. M. Stationery Office. London.

SIMEK, J. and HOLLIDAY, C. (2009). Rock shelter reveals rare art. Accessed 14 November 2009, at http://www.knoxnews.com/news/2009/jan/18/finding-an-aladdins-cave/

SKEATES, R (2009–10) Archaeological discoveries in the caves of Seulo, Central Sardinia. *The European Archaeologist* 32, 4–5.

SKEATES, R. (2012). "Caves in need of context: Prehistoric Sardinia." in K. A. Bergsvik and R. Skeates *Caves in Context: The economical, social and ritual importance of caves and rockshelters.* Oxbow Books, Oxford.

THURNAM, J (1861). Examination of a chambered long barrow at West Kennet, Wiltshire. *Archaeologia* vol. 38, 405-421.

WHITTLE, A., BAYLISS, A. and HEALY, F. (2008). The timing and tempo of change: Examples from the Fourth Millennium cal. BC. in Southern England. *Cambridge Arch. Journal* 18:1, 65-70.

Prehistoric Occupation and Use of a Cave near Seulo in Central Sardinia: 'The Oval Room' at 'Is Janas'

M. Giuseppina Gradoli

COMET – Valorizzazione Risorse Territoriali and ISSEP Sardinia, Via dei Colombi, 31, 09126 Cagliari

G. Terence Meaden

Oxford University, Department of Continuing Education (Archaeology) and Kellogg College, 62 Banbury Road, Oxford, England, OX2 6PN

***Abstract**: The Is Janas cave in the Barbagia di Seulo landscape of mountainous Central Sardinia has a grand primary branch with secondary chambers having archaeological deposits of major importance and simple fingerprint art. One branch, which embraces an auxiliary cavern in the shape of a roundish/oval room reached along a lateral passage with an awkward 'squeeze entrance', has been overlooked by previous archaeologists. Its plentiful, undisturbed deposits place its use in the Neolithic Age and Bronze Age. There are multiple hearths, deep ash, decorated pottery, and human bone and animal bone fragments. The oldest potsherds with dotted decoration and triangles are Neolithic of the Bonu Ighinu and Ozieri Cultures. Decorative marks on this pottery are matched to the tools that made them.*

***Key words**: Middle-Late Neolithic, Sardinia, Seulo, Bonu Ighinu Culture, Is Janas, 'decorated pottery with marking tool', sacred banquets.*

***Riassunto**: La Grotta Is Janas, situata nella regione montuosa denominata Barbagia di Seulo, Sardegna centrale, è costituita da un ramo principale con numerose camere ricche di frammenti di ceramiche, utensili, ossa animali ed arte rupestre. Un braccio secondario della grotta, accessibile attraverso uno stretto cunicolo, immette nella stanza tonda, così chiamata per la sua forma pressochè circolare. Quest'ambiente si è conservato completamente indistrurbato con i focolari lungo le pareti ed al centro della stanza, ancora in posto con i resti di pasto e gli utensili usati per la sua preparazione. Due pezzi di ceramica con decorazione puntiforme all'interno di campi triangolari sembrano appartenere alla Cultura di Bonu Ighinu, Neolitico medio. Con l'aiuto delle scienze forensi alcune decorazioni sono state comparate con gli utensili che le hanno prodotte.*

***Paroli chiave**: Neolitico, Sardegna, Seulo, Is Janas, 'utensili e ceramiche decorate', banchetti sacri.*

Introduction

This is the third summary about recent discoveries in the mountainous region of the Barbagia di Seulo in central-southern Sardinia. The main branch of 'Is Janas' is a fine show cave with huge stalactites and stalagmites and secondary chambers have bones, burnt ceramic sherds and some finger-mark rock art (Gradoli et al. 2005). Along one branch, the so-called 'Round Room' or 'Oval Room' is accessible by negotiating a squeeze tunnel, and there are two additional small chambers beyond where a 0.60m step permits entry into a space which has remained undisturbed since antiquity. There are several hearths, decorated pottery and numerous bone fragments. Potsherds with dotted decoration and a triangular zone are likely to be of the Bonu Ighinu Neolithic Culture. Shorter summaries of the initial inquiries have been published (Gradoli and Meaden (2012) and Skeates (2012)). See Gradoli and Meaden (2012a, 2012b) for current reports on other Neolithic discoveries in the Seulo region.

The round room of 'Is Janas'

The Is Janas limestone cave is located in the inner part of the Forest of Addolì. Inside the grand cave there is a secondary branch that can be entered along a narrow side passage by which is reached, after several metres and a tight entrance (Fig. 1), a further space 20m in diameter. This is the 'Round Room' where hearths were found (Fig. 2). The room's floor-surface is covered unevenly with a great quantity of stones, and with regions of hearths completely burnt together with pottery sherds and lithic tools of both 'Cultura di Bonu Ighinu' and 'Cultura di Ozieri', i.e. the Middle to Later Neolithic locally. Scattered all around are numerous animal bones (Fig. 3).

A few bones are carved (Fig. 4). Some of the sea-shells are shown in Fig. 4 and 5. One shell has red ochre on its external surface. In Fig. 6 some sherds from a hearth (shown in Fig. 2) can be seen where the other parts and the

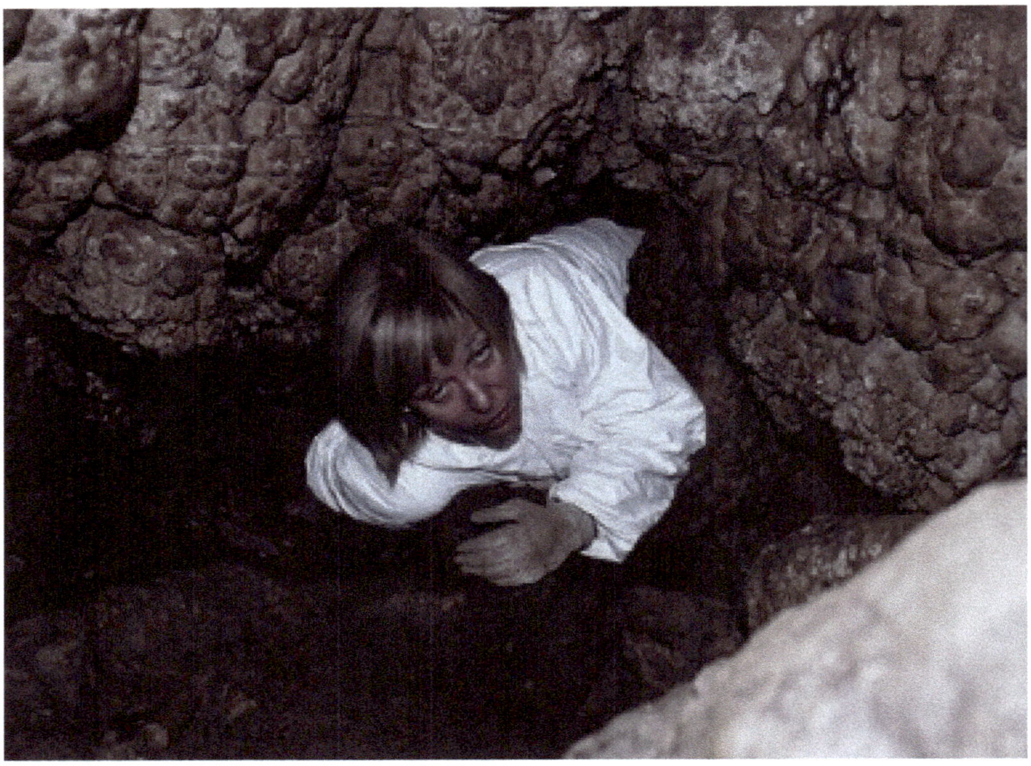

Fig. 1. Point of squeeze entry into the Round Room inside Is Janas

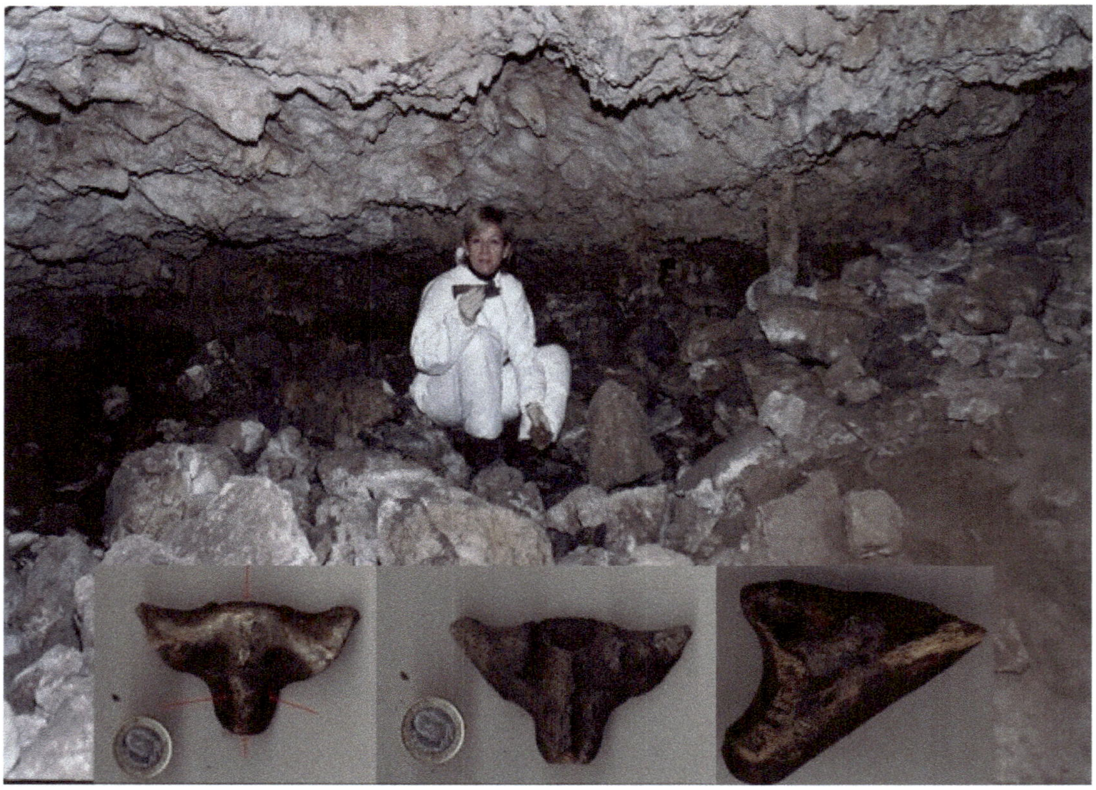

Fig. 2. Hearths in the cavern called the Round Room or Oval Room with pottery and bones.

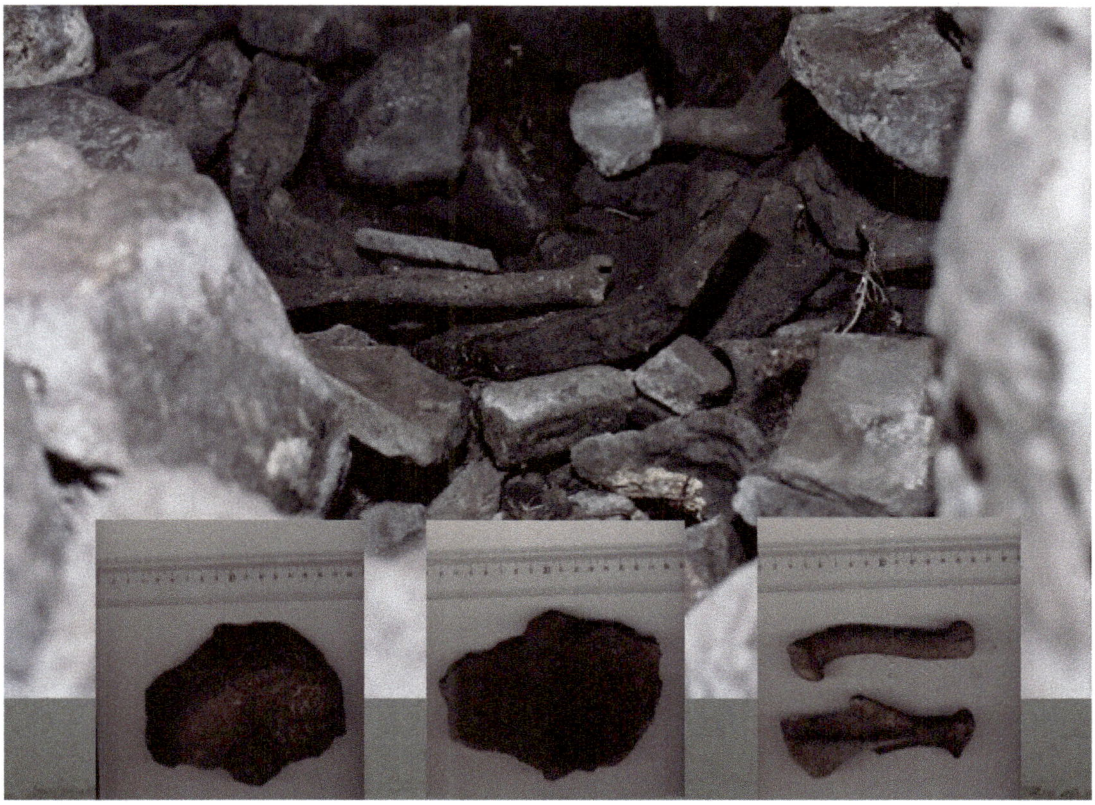

Fig. 3. Bones and pottery from the Round Room

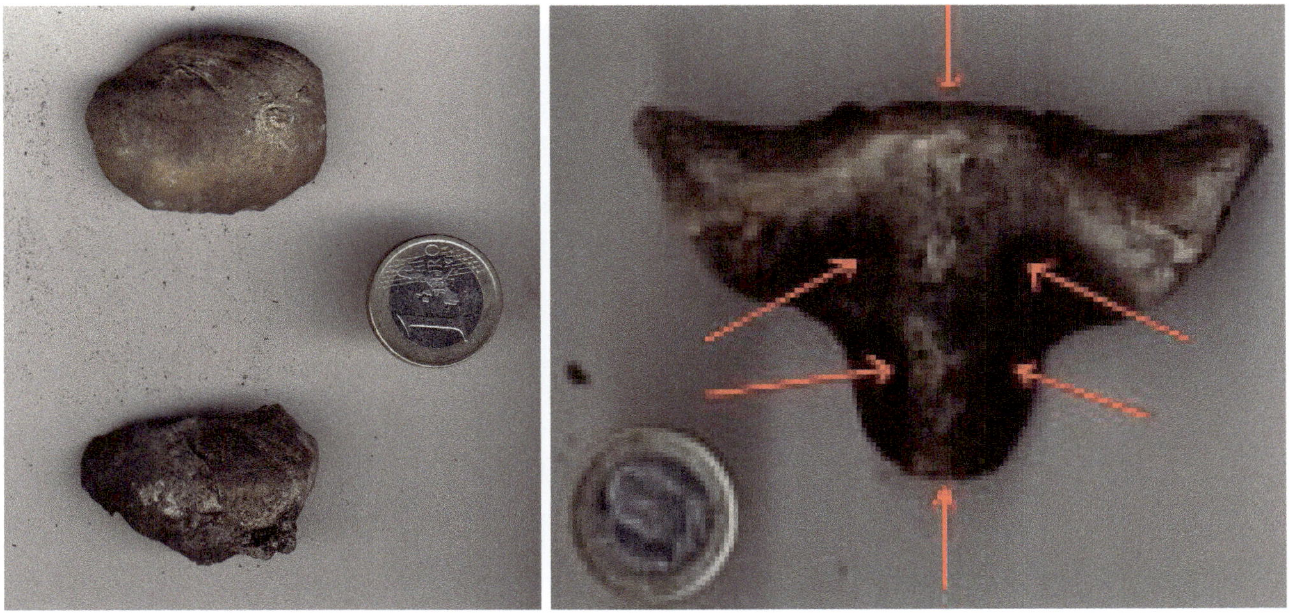

Fig. 4. Marine and terrestrial shells (left) and an intentionally smoothed animal bone (right). The holes in the tail bone are natural.

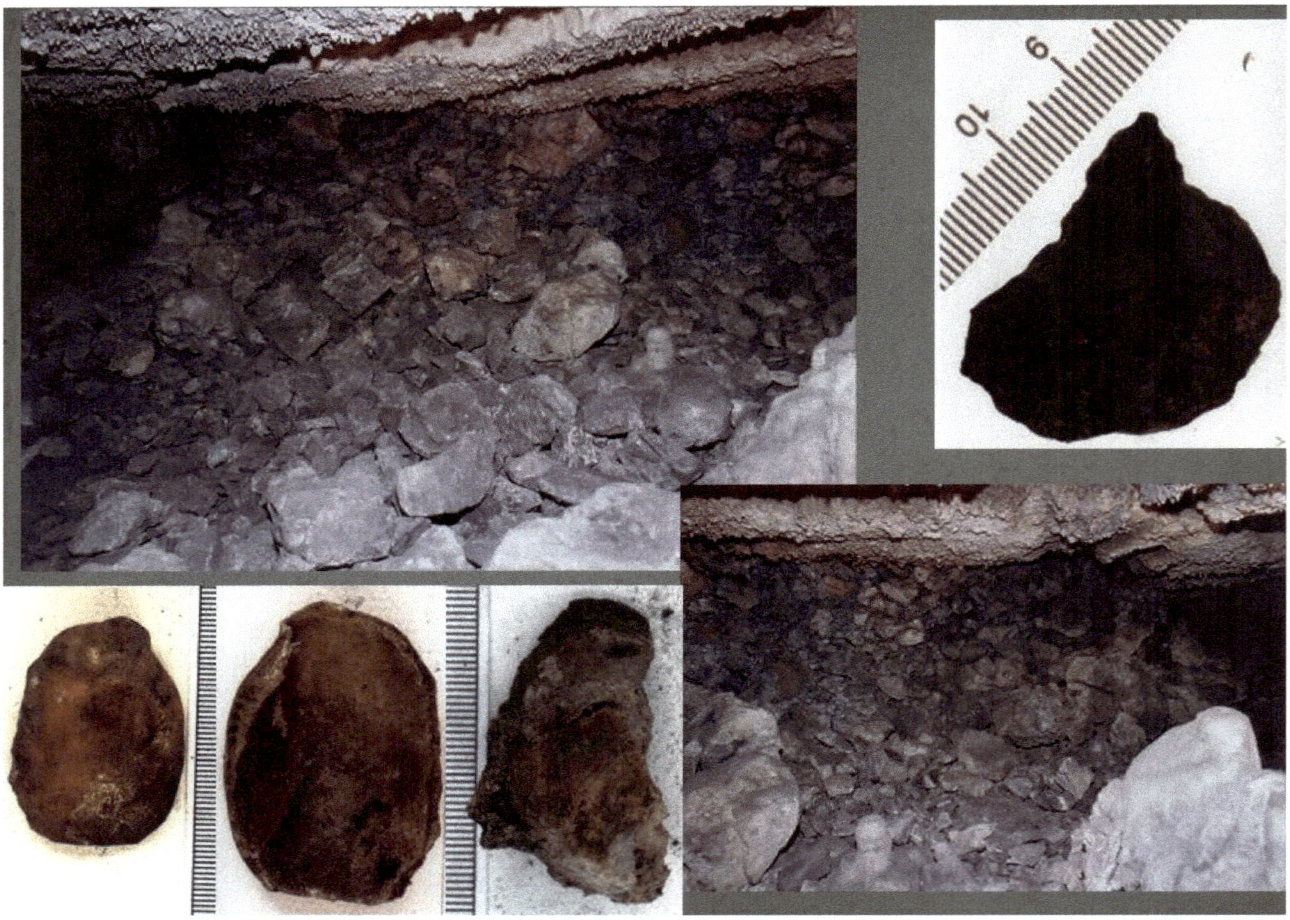

Fig. 5. A hearth (two views) and marine shells. One shell (at the left) is coloured with red ochre.

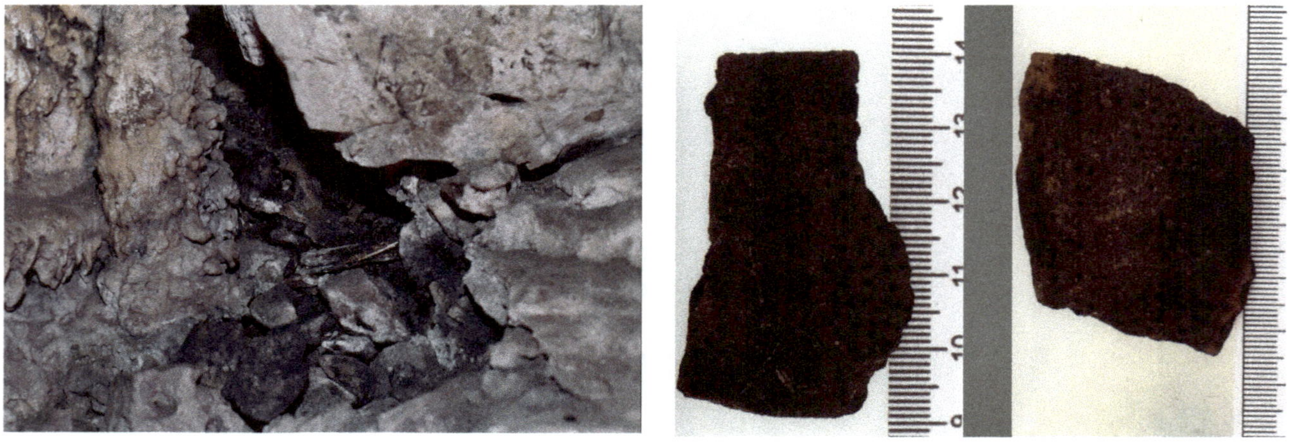

Fig. 6. The hearth with the 'Bonu Ighinu' Middle Neolithic pottery inside.

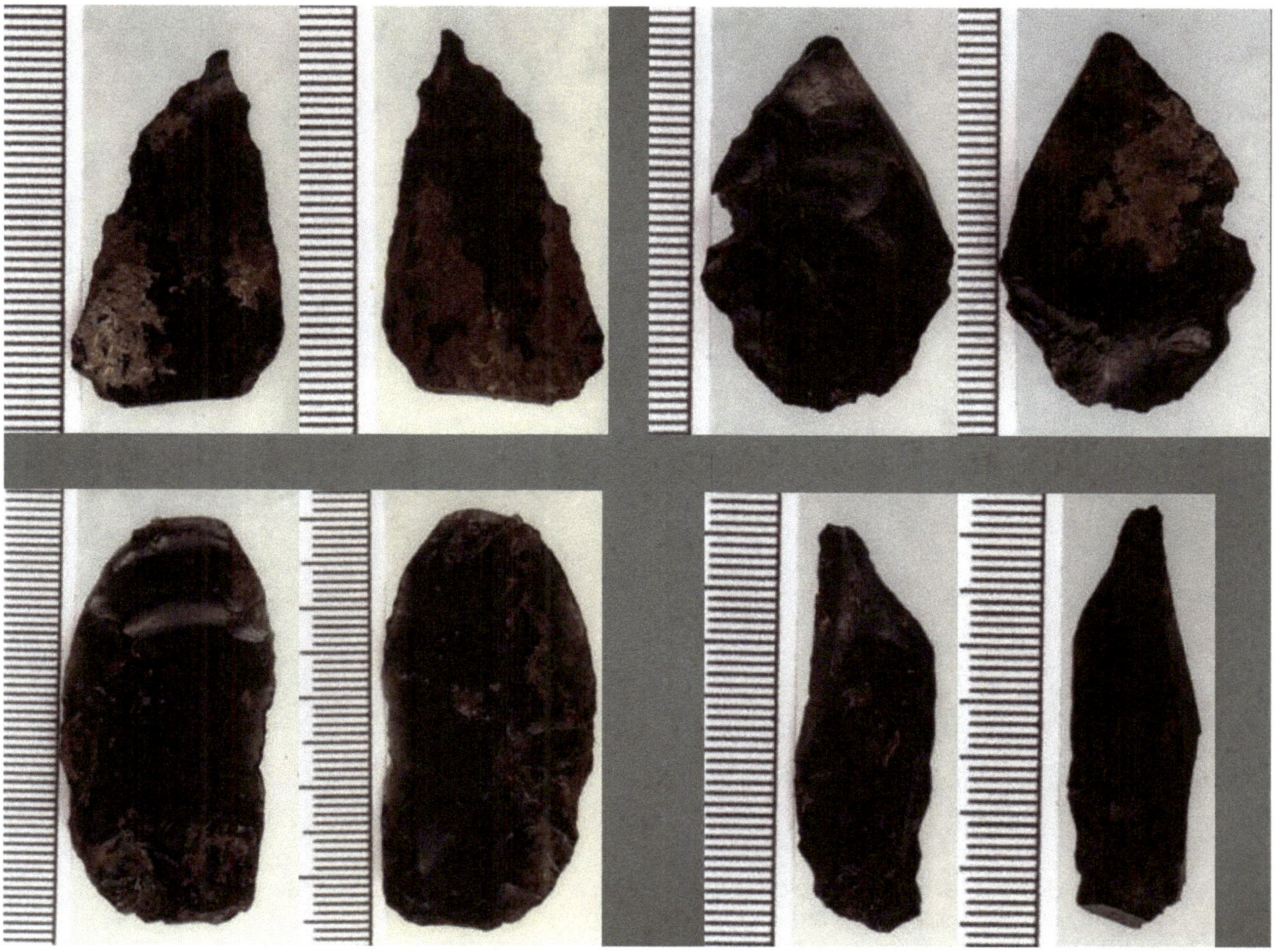

Fig. 7. Four well-worked obsidian flake tools.

handle are still *in situ*. There is charred, cooked food stuck to the base of the pot.

Figs. 7 and 8 continue with photographs of four black obsidian flakes, a flint blade and a pointed tool made of bone.

In Fig. 9, with the help of the Forensic Studies in 2008, the impressed sherds were matched with the tool that had marked them. The bone point (Fig. 8) had patterned the pottery while its truncated part (opposite the point) had impressed another sherd.

Additional smaller rooms. From the Round/Oval Room, after descending a 60cm step, there are two small round spaces with burned animal bones. A thick layer of ashes, partially calcified with bones, plant remains and impressed potsherds of Ozieri Culture (Later Neolithic) were present too (Fig. 10).

It seems that the Round Room was a special place where sacred banquets were held. This is why we find the cave floor-surface burnt up and with many hearths and places with animal bones, potsherds, and lithic and bone tools. No human bone was found even after the excavation of the two one-metre squares in 2009. Excavations, led by Dr. Skeates—who we informed of our 2005-2008 discoveries at the EAA Meeting in Malta in September 2008—began in Summer 2009 with Dr. Gradoli.

Concluding remarks

The numerous caves of the carbonitic plateaux of the Barbagia di Seulo seem to have been used for primary burials of the dead in Neolithic times, as well as serving as places where certain 'passage rites' were performed or where shrines were dedicated probably to the Earth Mother. Sometimes the dead were buried and then covered with stones; at other times they were laid on the floor of the cave, or, as noted in one remarkable case (Gradoli and Meaden 2012b), were enclosed and preserved in a *stalactite natural coffin* created by calcitic water-drip, possibly deliberately. Secondary burials were also found in little chambers and holes in the rock dating from Neolithic times, although this practice may have been continued into Bronze Age and later periods. All told, the Barbagia di Seulo constitutes a rediscovered ritual, sacred landscape dating from Middle Neolithic times and which will warrant much expert study in the coming years.

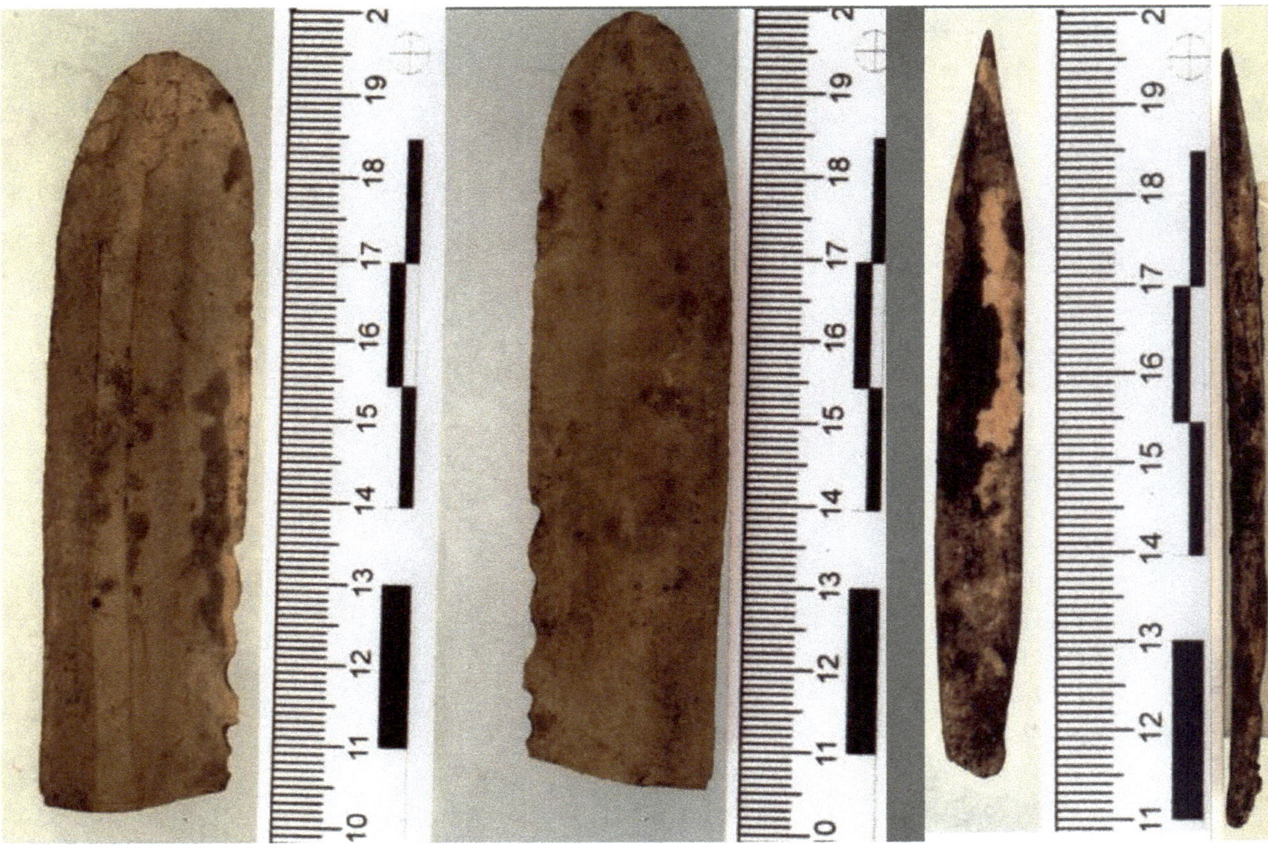

Fig. 8. An Ozieri Culture (Later Neolithic) flint blade and a 'Bonu Ighinu' (Middle Neolithic) bone point.

The position and use of the 'Is Janas' roundish oval cavern within the overall Neolithic scenario for Barbagia di Seulo needs to be referenced to the information given in the preceding two papers (Gradoli and Meaden 2012a, 2012b). The cave is not alone in the geological landscape with its various aspects involving surface features, woodland, trees, soils, stones, rock-faces or cliffs, water, underground qualities, and even climate and weather variations. Every site has innate potential for the sacred. Caves have humidity and, apart from the eerie splash of dripping water, a memorable quietness and a special ambiance and unforgettable atmosphere.

Acknowledgments: We thank Giuseppe Farci for many of the photographs used in these three papers.

References

BERGSVIK, K.A. and SKEATES, R. (2011). *Caves in Context: The economical, social and ritual importance of caves and rockshelters.* Oxbow Books, Oxford.

GRADOLI, M.G. et al. (2005). Il binomio uomo-territorio nella Barbagia di Seulo (Sardegna centrale): prime segnalazioni di pitture parietali ed incisioni in grotte carsiche. Atti del XVI Convegno degli Antropologi Italiani dedicato al "Processo di Umanizzazione". Genova (29-31 ottobre 2005). Edicolors, Genova.

GRADOLI. M.G. and MEADEN, G.T. (2012). "Discovery of prehistoric sites in caves and rockshelters in the Barbagia di Seulo, South-Central Sardinia", in K. A. Bergsvik and R. Skeates *Caves in Context: The economical, social and ritual importance of caves and rockshelters.* 188-197. Oxbow Books, Oxford.

GRADOLI, M.G. and MEADEN, G.T. (2012a). Sacred sites and symbolism in the Neolithic landscape of Barbagia di Seulo, Central Sardinia. In Meaden, G.T. (2012) *Archaeology of Mother Earth Sites and Sanctuaries through the Ages: Rethinking symbols and images, art and artefacts from history and prehistory.* British Archaeological Reports, Oxford.

GRADOLI, M.G. and MEADEN, G.T. (2012b). Neolithic art and artefacts in caves near Seulo in Central Sardinia: 'Grutta I De Longu Fresu'. In Meaden, G. T. (2012) *Archaeology of Mother Earth Sites and Sanctuaries through the Ages: Rethinking symbols and images, art and artefacts from history and prehistory.* British Archaeological Reports, Oxford.

SKEATES, R (2009–10) Archaeological discoveries in the caves of Seulo, Central Sardinia. *The European Archaeologist* 32, 4–5.

SKEATES, R. (2012). "Caves in need of context: Prehistoric Sardinia." in K. A. Bergsvik and R. Skeates *Caves in Context: The economical, social and ritual importance of caves and rockshelters.* Oxbow Books, Oxford.

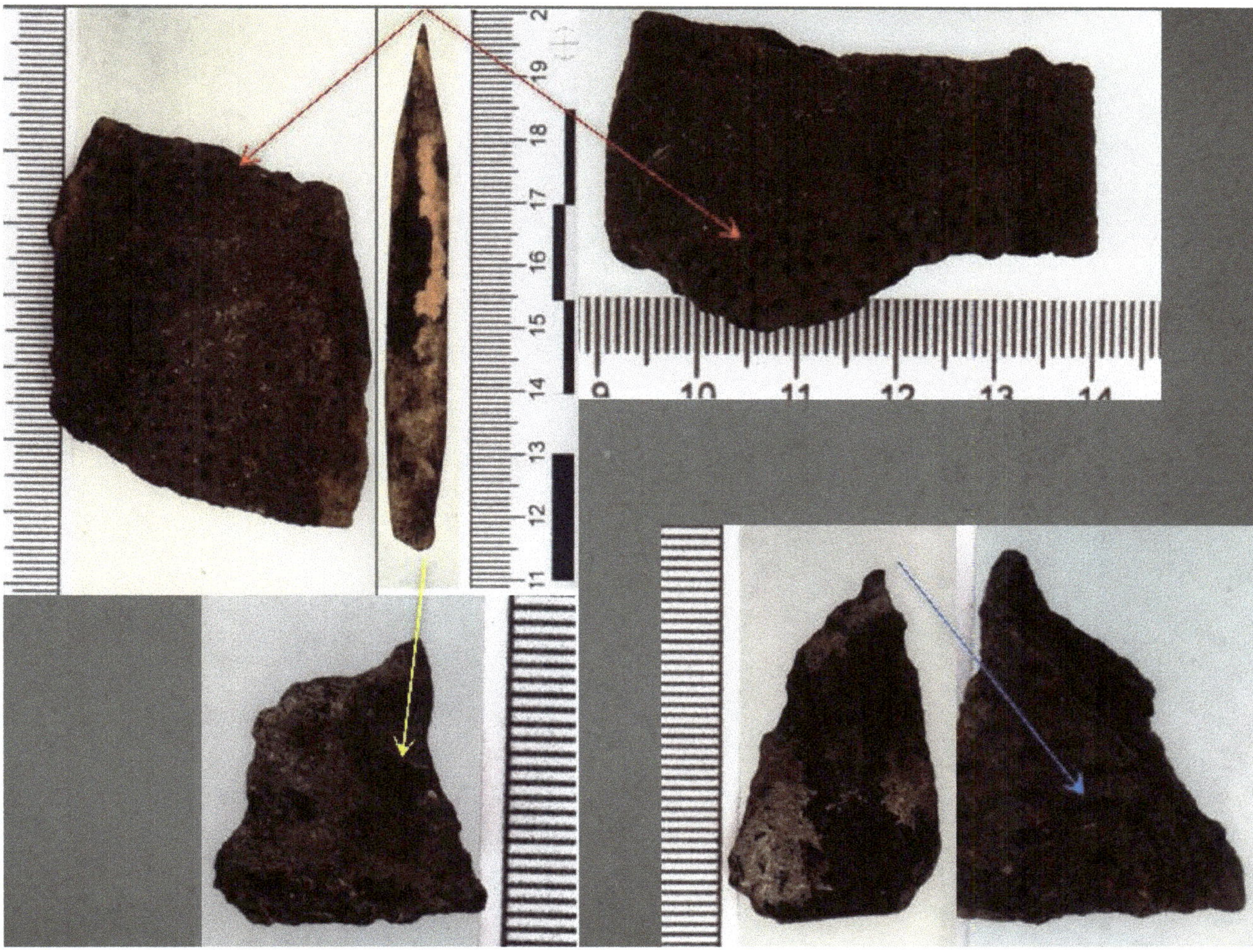

Fig. 9. Potsherds that have been marked by the bone tool indicated.

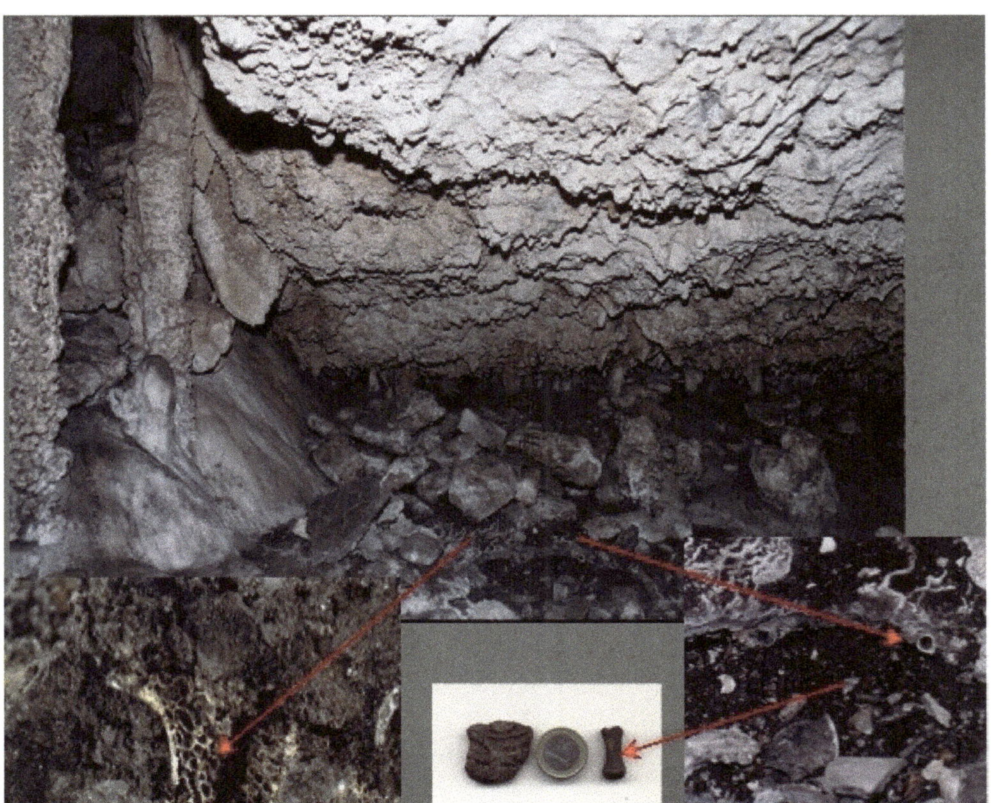

Fig. 10. Part of the smaller inferior room at Is Janas Cave.

THE NEOLITHIC MONUMENT OF NEWGRANGE IN IRELAND: A COSMIC WOMB?

Kate Prendergast

Oxford University, Department of Continuing Education

Abstract: *This paper argues that the Neolithic monument of Newgrange, in common with comparable monuments known as passage-graves, functioned to facilitate womb-like ritual experiences and birth-based cosmological beliefs. It explores the evidence for the design, material deposits, astronomy, rock art and associated myth at Newgrange to suggest the myriad ways that birth-based ritual and cosmology are invoked at the site, and it locates this evidence in the context of the transition to agriculture with which such monuments were associated.*

Key words: *Neolithic, Newgrange, monument, womb, womb-like, ritual, astronomy, winter solstice, re-birth, ancestors.*

Introduction

The passage-grave (or passage-mound) of Newgrange (Fig.1), along with its neighbouring passage-graves Knowth and Dowth, are among the most outstanding Neolithic monuments to survive in Britain and Ireland. Situated by the Boyne River in County Meath in eastern Ireland, each monument contains a long passage and chamber constructed from stone and covered by a large circular earth and stone cairn (Knowth and Dowth each have two passages and chambers). The excavations and reconstruction of Newgrange and Knowth have revealed many extraordinary things about these monuments (O' Kelly 1982; Eogan 1986). These include dating their construction and use to the later Neolithic (c3200-2500 BC), documenting the extensive and highly accomplished rock art inscribed on the megaliths at the sites, and establishing that at Newgrange there is a precise astronomical alignment built into the monument. Further work has demonstrated there are astronomical alignments at Knowth and Dowth and persuasively linked the rock art and astronomy to suggest the myriad ways that the sites may have functioned as cultural and ritual centres (see for example, Brennan 1983, Moroney 1999, Cooney 2000).

It is now fully accepted that the Boyne valley sites represent the culmination of the Irish passage-grave tradition. In areas such as Brittany, passage-graves had existed since at least 4500 BC (Sheridan 1986). It is possible that Breton monumental styles influenced the development of passage-graves in Ireland, which began to be constructed around 3800 BC and reached their largest and most elaborate form at the Boyne sites c.3200 BC (Berg 1995). Moreover, it has been argued that these large prominent sites had a seminal influence on the emergence of open-air ritual centres such as henges and stone circles in the far north and west of Britain and eastern Ireland around 3000 BC (Bradley 1998). In other words, the monumental and ritual tradition associated with passage-graves in Ireland was in existence for over a millennium and as its zenith, the Boyne sites influenced the developmental path that monuments took in the British Isles during the third millennium BC. Passage-graves, therefore, were clearly places of real power, and embodied key cultural and ritual concerns of the people who built and used them.

This paper argues that one of the primary sources of power embodied in monuments such as Newgrange was the way in which they functioned in a ritual sense like a womb, and that rituals related to beliefs about birth were key for prehistoric peoples in making sense of their world while engaged in a long-term transition from hunting and gathering to domesticated food production.

In order to present detailed evidence for the ways in which developed passage-graves may have been constructed and used to facilitate 'womb-like' ritual experiences, this paper focuses on the Newgrange monument. It explores the evidence for architectural design, astronomical alignment, rock art and associated mythology at Newgrange to suggest how—and perhaps why—such ritual experiences may have been so important in the Neolithic.

The design of Newgrange

Like Knowth and Dowth, the circular mound or cairn covering the passage and chamber at Newgrange is over an acre of ground in size and approximately 10m in height (Eogan 1986:44-5). It covers a long passage and a chamber built from orthostats and lintels and topped with a cap-stone. The cairn has a kerb constructed of massive contiguous slabs surrounding the external base of the mound. The kerbstones, and the passage and chamber orthostats and lintels are all decorated with rock-art. The inscribed motifs include spirals, wavy lines, concentric circles, dots, zigzags and chevrons. The entrance at Newgrange is demarcated

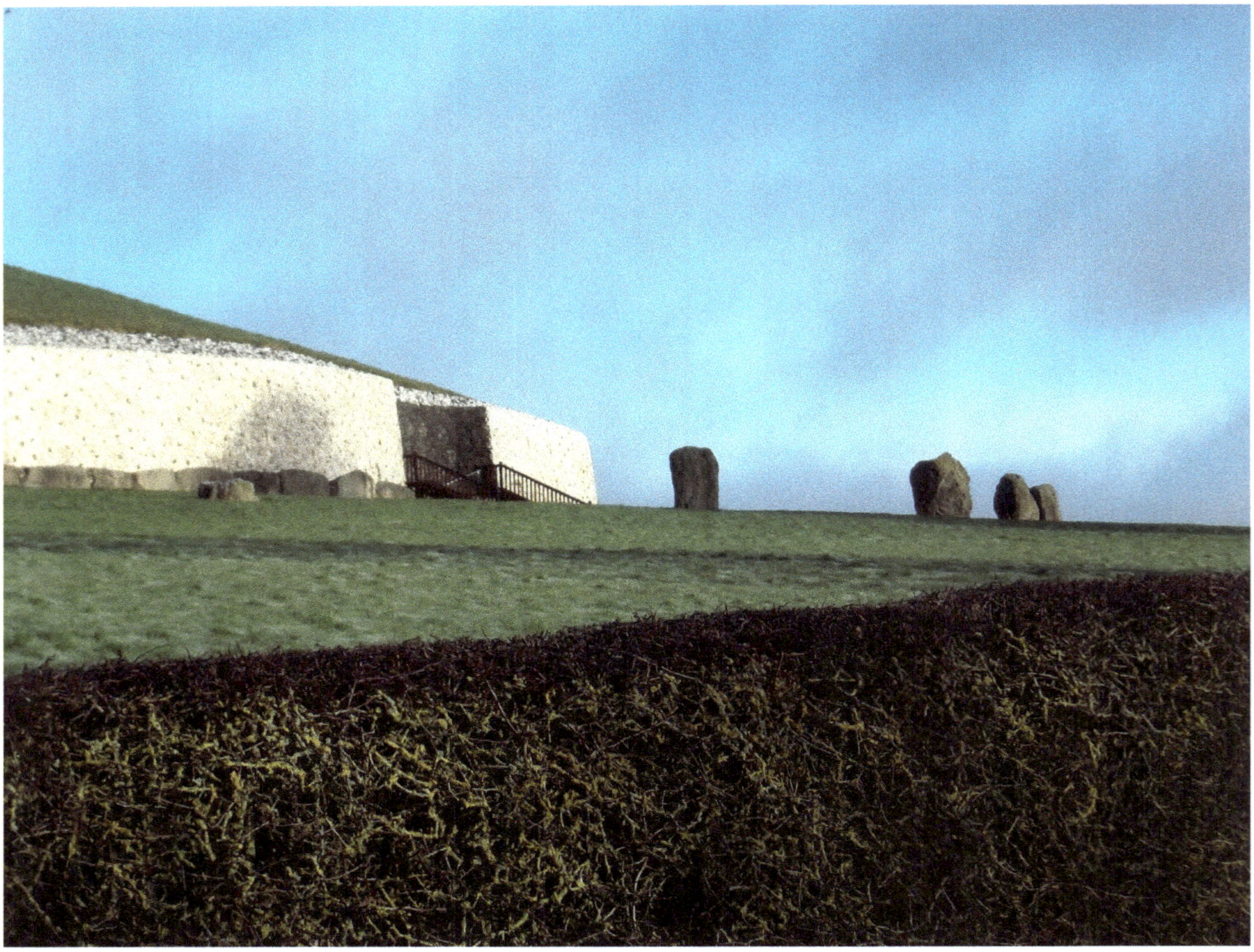

Fig. 1. The reconstructed façade of Newgrange showing also some standing stones of the concentric stone circle.

by a huge decorated kerbstone, divided by a vertical line that also marks the centre of the passage entrance. Directly above the entrance is a further rectangular opening into the passage—known as a 'roof-box'. Quartz features prominently in the covering of the mound at Newgrange, which gleams when it catches the light of the sun, and the remains of a circle of large standing stones surrounds the monument.

It is the architectural properties of the passage and chamber that closely resemble a womb-like space. At Newgrange, the passage runs SE-NW. It is 24m long and leads to a cruciform chamber with three recesses. The path of the passage follows a fairly steep incline: the chamber is up to two metres higher than the passage entrance. The low narrow passage bends slightly. The combined effect of these features means that no natural light penetrates the chamber, except on the dawn of the winter solstice – a highly significant astronomical alignment built into the construction (see below). The only way it is possible to see the entrance from the chamber is to lie down by the back recess, where the merest chink of light from the entrance is just visible at floor level.

The chamber therefore was designed to be in total darkness, except for about 17 minutes once a year. The corbelling of the chamber roof is so well-finished that the chamber is completely dry. No light, water or sound penetrates this inner sanctum, buried deep in a huge earthen mound at the end of a long and narrow passage. It would be difficult to think of a space more womb-like in its construction, and it is almost certain to be deliberate.

Astronomy at Newgrange: the winter solstice rising sun

The fact that Newgrange was designed according to highly specific principles, in order to produce particular effects, is nowhere more apparent than in its astronomical alignment. Traditional local stories had long suggested that the sun entered the passage and chamber at Newgrange, while the south-easterly orientation of the passage indicated an alignment with the rising sun around the winter solstice. During the early stages of excavations by Prof. Michael O' Kelly, it was decided to investigate this scientifically. In 1969, the excavators, crouched in the passage at dawn on winter solstice, reported that:

At exactly 8.54 hours GMT the top edge of the ball of the sun appeared above the local horizon and at 8.58 hours, the

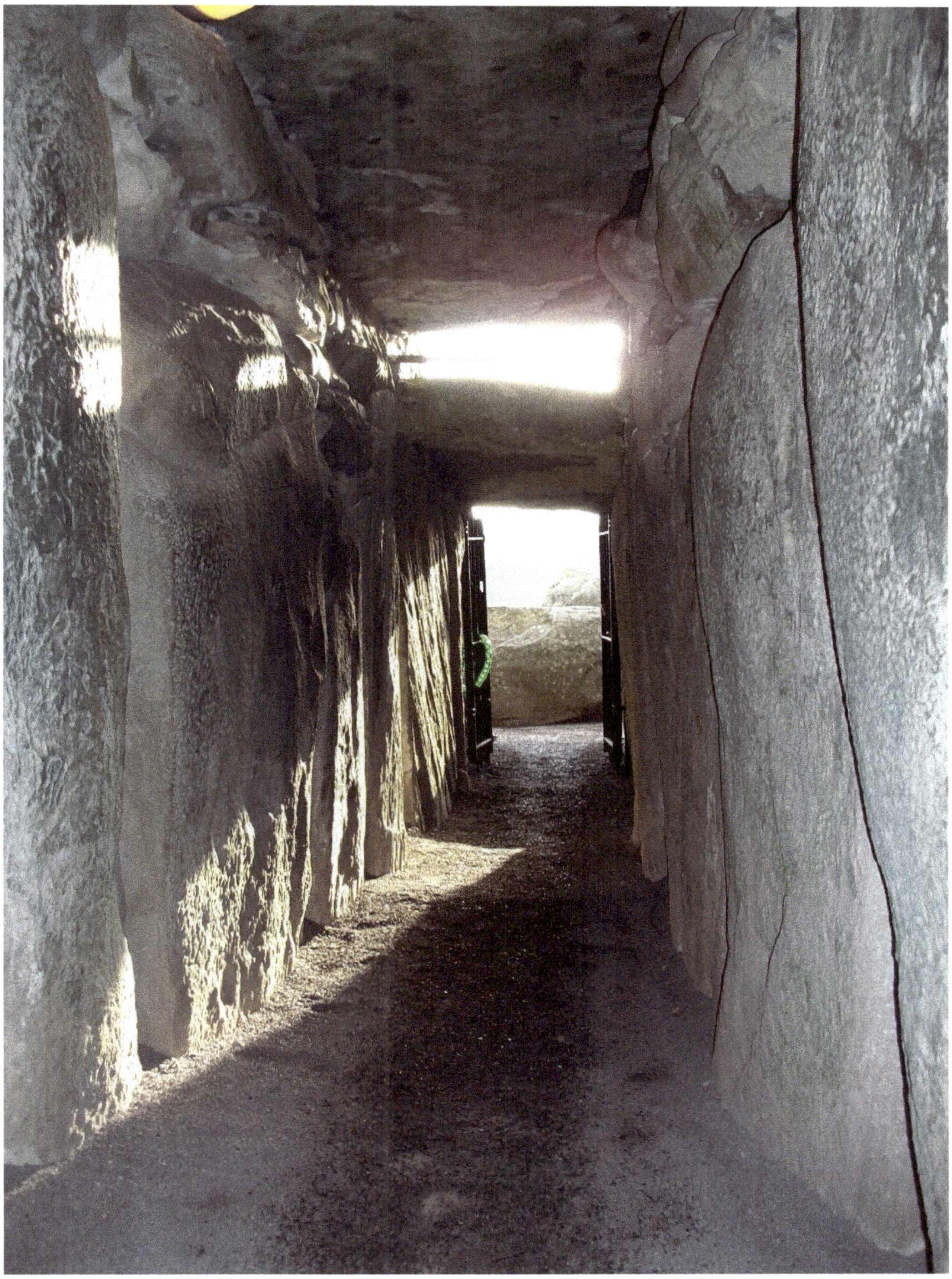

Fig. 2. The entrance part of the gallery that is the axis of the monument.

first pencil of direct sunlight shone through the roof-box and along the passage to reach across the tomb chamber floor as far as the front edge of the basin stone in the end recess. As the thin line of light widened to a 17cm-band and swung across the chamber floor, the tomb was dramatically illuminated and various details of the side and end recesses could be clearly seen in the light reflected from the floor. At 9.09 hours, the 17-cm band of light began to narrow again and at exactly 9.15 hours, the direct beam was cut off from the tomb. For 17 minutes therefore, at sunrise on the shortest day of the year, direct sunlight can enter Newgrange, not through the doorway, but through the specially contrived slit which lies under the roof-box at the outer end of the passage roof (O' Kelly 1982: 124; Fig.2).

Further research has demonstrated just how precise this alignment is. Patrick (1974) has proved that the winter solstice orientation was operative when Newgrange was constructed, and therefore is an original, central and permanent feature of the monument. O' Brien (1988) has shown that the chamber and passage are sophisticated and complex constructions, designed to maximize the accuracy and length of the beam of light coming into the chamber. His research indicates that at the time of construction, the beam of light entering the passage at Newgrange was so precisely framed by the roof-box that it could be used to calculate the day of the solstice itself. In short, Newgrange was precisely aligned on the winter solstice rising sun in such a way it could be used as a 'clock' to calculate the exact moment of the solstice.

The design of Newgrange: facilitating ritual experience

It is when we put together the construction of Newgrange, including its astronomical properties, with the experiences engendered by such a design, that we can begin to appreciate just how powerful a place Newgrange was (and still is) for the facilitation of particular ritual experiences.

It is possible, for example, that Newgrange was a site for initiation. Such rituals are suggested by the experience of being in the chamber before and during the entry of the winter solstice sun (Prendergast 2010). Waiting for the sun to enter the passage and chamber before dawn is to wait in a buried, dark, dry, silent circular space. It is like being blindfold and bound, in a liminal space, awaiting an enlightening event—like waiting to be born. When the light enters, it is as a single pencil of red light, filling the dark chamber with intimations of the world outside—the 'opening' of the birth canal for a baby to take its passage to the external world. Leaving the chamber is as if to be born, creeping hunched down the passage, away from enclosure, darkness and stillness, to light, sound, movement and sky.

Moreover, this event happens at a very significant time in the seasonal and celestial cycle. Mid-winter is at the heart of the season of death and decay. It heralds the shortest day and longest night, and the slow turn of the earth towards spring—towards new light, warmth and life. As such, it is a crucial moment to record in terms of seasonal change. At a ritual level, it could also be interpreted and experienced in terms of death and birth, because it is, quite simply, the astronomical moment when the earth 'dies' and is 'born'.

Death and rebirth

The sense that a 'birth' occurs both ritually and cosmically at Newgrange on the winter solstice is strengthened when we consider the deposition of human remains in the chambers of the monument.

Cremated and un-burnt human bone deposits were discovered in Newgrange in the SW and NE chamber recesses (O' Kelly 1982: 102-7). They were closely associated with the circular granite basins found in the recesses, leading some to speculate that the carved basins held burial deposits. The presence of human remains in passage-graves is what has led such monuments to be classified as 'graves'. But it is more likely that they were used in complex ways, and the deposition of human remains was one activity among many practiced at the sites. If this is the case, it suggests such rituals, and associated beliefs, involved the living *and* the dead.

Alongside the invocation of birth, perhaps in an initiatory context, this evidence allows us to consider the possibility of the invocation of the more complex idea of 're-birth'. It may be that the builders and users of Newgrange believed that, along with a ritual birth experienced by the living, and the cosmic birth that happens on the winter solstice, the dead – or the ancestors – also experienced a re-birth at this time. If this was the case, it brings together the living community, the dead and the natural world in a powerful set of inter-related beliefs.

Rock art

The rock art at the Boyne valley sites is highly accomplished, profuse—together it represents the largest collection of megalithic rock art in Europe—and emphasises particular abstract designs, presented singly or in combination: spirals, circles, crescents, zigzags and wavy lines.

Martin Brennan has perhaps done the most to explore the meanings embodied in this art and to make connections between the art and astronomy at all three sites (1983:127-205). Several stones suggest a very strong connection between the rock art and astronomy. Stone SW22 –a kerbstone at Knowth—Is one: Brennan has convincingly argued it depicts the monthly cycle of the moon—and possibly even more complex lunar cycles such as the Metonic cycle (1983:144).

This paper focuses on the prominent and recurring image of the triple spiral at Newgrange. The triple spiral is the central image at Newgrange. It appears on the entrance kerbstone (Fig.3). It appears again about halfway up the passage on stone L19 (Fig.4). It found again on the right hand wall of the back recess in the inner chamber (Fig.5).

The simple fact that the triple spiral has been placed so centrally at Newgrange would suggest a connection with the winter solstice alignment. This would appear to be confirmed by the fact that each image has been located to directly receive the light of the rising sun as it enters the monument at winter solstice (Brennan 1983:80-1).

A spiral is a natural symbol for the movement of the sun. The movement of the sun in the sky over a year can be depicted by a spiral and the curvature of shadows of the sun over a year form a double spiral. In summer the spiral is clockwise, in winter, anti-clockwise. The three sets of triple spirals, especially those on the kerbstone and in the back recess wall, display an interest in the relationship between three and six (Brennan 1983: 193-194). Three and six are basic counting units and have an integral relationship to astronomy, because they are the number of months between the cross-quarter days and solstices/equinoxes and between the solstices and equinoxes respectively. Moreover, some of these individual spirals are 'double', i.e. each one has

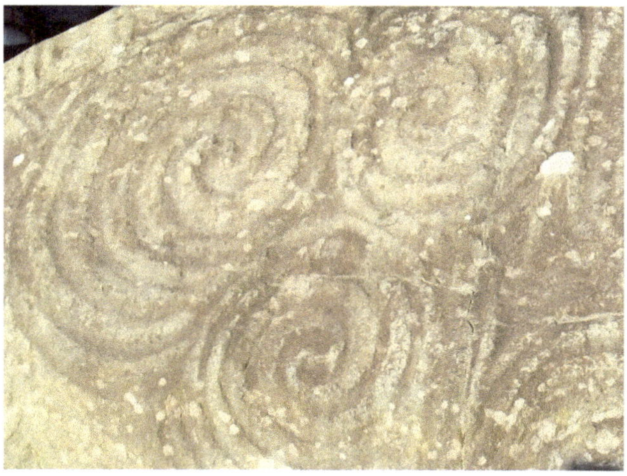

Fig. 3. A stone carved with triple spirals

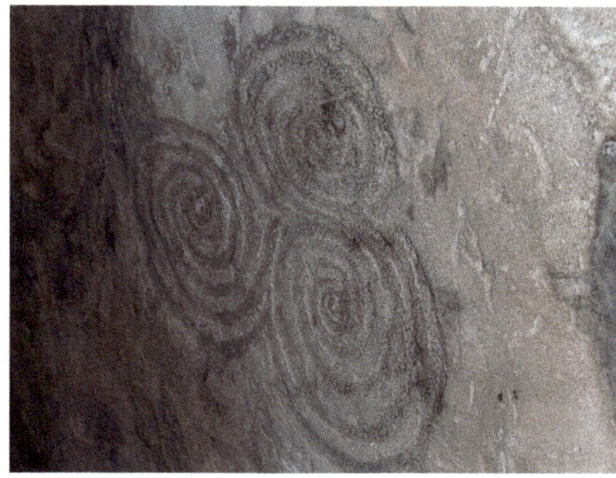

Fig. 5. The triple spiral carved into the end chamber of Newgrange

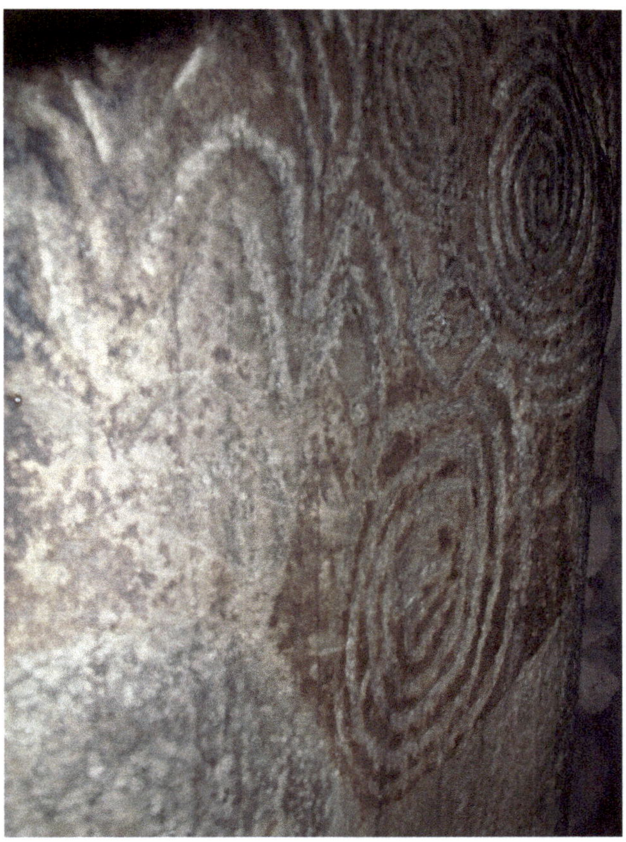

Fig. 4. A side stone of the gallery carved with triple spirals.

two sets of spirals or counts, and where they meet in the middle could be seen to be a representation of alpha and omega—the beginning and end point of each journey that the spiral represents.

As such, they may represent the interplay between counts of three, six, nine and twelve. Such interplays include the number of months in a season (three), the number of months in a solar year (twelve), and the number of months in a human gestation period (nine). Three sets of three spirals would add weight to the importance of the number nine. Could these images have been used to enable women to calculate the timing of pregnancy and birth according to the lunar and solar cycles? (Refer to Meaden (1997: 134-144) and to Fig.6 for sexual associations in the way the phallic shadow of the standing stone at Newgrange is near to 'penetrating' the passage entrance at sunrise on the winter solstice—as it certainly does at Stonehenge and Avebury at the summer solstice).

The basic unit underpinning such measurements is three. We see three reproduced in the number of spirals in each set of triple spirals, the number of triple spiral images, the number of recesses in the chamber and the number of large passage-graves at the site. Three is also the number that underpins the Trinity. Pagan conceptions of the Trinity have a female aspect: the maiden, mother and crone (Matthews 1989). These three ages relate directly to birth, since they are the basic way we calculate a life, and they interconnect with reproduction – the fertile age of the mother when women have children. In male aspects of the Trinity, more emphasis is placed on the new-born child as one of its elements, but in either case, fertility and birth is a central part of the concept.

The triple spiral or Triskele became a prominent symbol in Celtic Christian art, notably in the Book of Kells held at the nearby Abbey of Kells. It was used to represent the Christian Trinity, but this raises the intriguing possibility

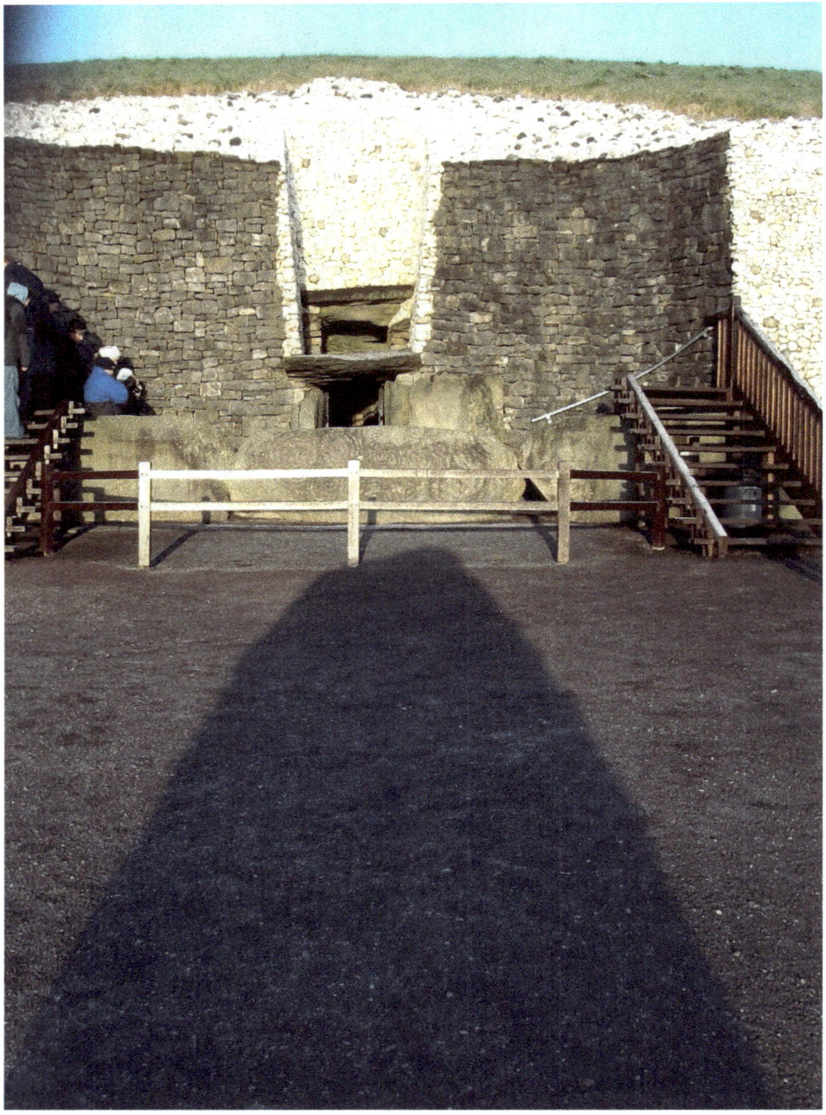

Fig. 6. The shadow of a standing stone shortly after sunrise in midwinter week is angled along the axis of the monument and nearly touches the carved boulder recumbent at the entrance.

that the early Christian monks were directly drawing on older pagan conceptions and depictions of the Trinity. The fact that the Celtic pagan gods were connected to the Boyne valley sites is documented in the myths and stories captured in the Irish *Mythological Cycle*, first written down in the early Middle Ages (O' Kelly 1982:43). Remarkably, such stories also invoke the birth of the gods. In so doing, by association, they bring the human, natural and metaphysical elements invoked and orchestrated at Newgrange to new levels of interconnection and integration.

Myth

The *Mythological Cycle* in part concerns the Tuatha Dé Danann, god-like ancestors of the Irish. The Tuatha Dé Danann were descended from the goddess Danu and came down from the sky disguised as a supernatural race of wizards and magicians. The Bru Na Boinne—the 'houses' on the Boyne—are invoked in the literature as the residence of these gods, and as a magical place where no-one dies, where there are always festivities, and where magic fruit-trees and inexhaustible cauldrons ensure that no company ever went away unsatisfied. Such a place echoes the widespread descriptions of the 'otherworld', both in Dark Age and medieval oral Celtic traditions and in more recent oral traditions, recorded in Ireland as recently as the early twentieth century (Evans-Wentz 1911). In such descriptions, the 'otherworld' has a liminal status in much the same way as that suggested by the interior of Newgrange. It is a place that people can disappear to, often through 'doors' located in natural features of the landscape like hills and stones. It is not uncommon to have to spend a year in the otherworld, but it is equally uncommon not to return from it. In this sense, it truly is a place where 'no-one dies', since it is a place where beings go to and *return from*—in the same way that death may lead to re-birth at Newgrange.

In the myths, Newgrange is associated first with Elcmair who was married to Boand, the divinised personification of the Boyne river. Newgrange is sometimes known as 'Bru mna Elcmair', or the 'House of the woman of Elcmair'. They are then associated with 'Dagda', among the most prominent of the old Irish Gods. His name has been translated as 'good god', and he is also called 'Ruad Ro-fhessa', 'The Lord of Great Knowledge' (Brennan 1983:10-11).

Dagda gains possession of the Bru Na Boinne and achieves carnal union with Boand by utilising his mastery over time. He sends Eclmair on an errand which lasts for nine months, during which time Oengus, or 'Mac ind Oc'—'the Youthful Son'—Is conceived and born. Boand describes Oengus as the son who was begotten at break of day and born betwixt it and evening. His birth is associated with a magical lengthening of the day at Newgrange. Newgrange is most widely known in the ancient literature as 'Bru Mac ind Oc', or the 'Bru of Oengus', as Oengus takes over habitation of the mound from Dagda, by utilising his own power over time.

The imagery surrounding Oengus in these accounts would appear to confirm that in Irish oral traditions of the early Middle Ages, Newgrange was associated with birth at winter solstice—a place where a god was conceived and born. In these myths, the entry of the sun into the passage and chamber at dawn is the moment of conception (i.e. the marriage of the gods), and the subsequent day the product or offspring of this union (the birth of the cosmos). This reiterates both the evidence for a ritual birth (initiation) at the site and the logic of the Trinity suggested in the triple spiral, in which divinity procreates and adds to itself through a sacred birth. It is notable in this context that in Rome in the fourth century AD, Jesus' birth was synchronised with the winter solstice—a direct echo in the Christian Trinity of earlier pagan and prehistoric beliefs and practices revealed at Newgrange (Miller 1993:10-11).

Conclusion

This paper has explored evidence for the ways in which the Neolithic monument of Newgrange may have functioned to facilitate beliefs and rituals around an inter-connected set of motifs related to birth. These include a withdrawal into a womb-like interior that is echoed by stories of withdrawal into the otherworld, and a 'birth' or return down the passage to the world outside. The presence of the dead in the chamber recesses may indicate further beliefs in re-birth. Moreover, such events occur alongside the birth or re-birth of the cosmos on the shortest day of the year. Oral traditions indicate a direct relationship between pagan Celtic gods and the site, and these and the imagery of the triple spiral further invoke the idea of a pagan Trinity centred on the motif of a sacred birth. The triple spiral may also be a depiction of astronomical measurement and counting, and may include specific counts of the human gestation period. The triple spiral is later reclaimed by early Christians to represent the Trinity in Christian form, whose male god is also born out of divine union on the winter solstice.

Such a powerful set of associations raises some interesting questions. The first is the extent to which Neolithic beliefs and practices at such sites had a female—or more specifically a mother—focus. The second is why such beliefs continued to be echoed in later descriptions of the sites and indeed in later religious imagery, which in some cases appears to have directly drawn on earlier pagan associations between the Trinity, the winter solstice and divine birth.

It has been recognised for some time that passage-graves were an indigenous hunter-gatherer response to the encroaching influence of farming as it spread across northern and western Europe during the fifth millennium BC (e.g. Bradley 1998). While the onset of the monumental tradition in north-west Europe appears to have reproduced southern European Neolithic conceptions of the house in the long mound, passage-graves were a radical departure from this tradition inasmuch as they opened up the monument with the insertion of a passage and chamber. Unlike long mounds which were closed burial sites, passage-graves could also be used by the living, and in particular they facilitated a relationship between the living and the dead. Archaeologists such as Barrett have argued that this reflects indigenous views of the ancestors—of collective ideas of ancestry focused on a place in the landscape—in contrast to the closed conceptions of the house (and by implication lineage) associated with agricultural traditions (Barrett 1994). From the evidence explored in this paper, it would seem possible that the place where the ancestors could be accessed was also conceived as a collective womb—even perhaps as a representation of a common female ancestor: a hunter-gatherer 'Eve'.

Neolithic groups in southern Europe and the Near East also drew on explicit images of female fertility and reproduction. Maria Gimbutas and others have documented the extensive use of images of mother figures among such groups; images that themselves drew on hunter-gatherer representations of female figures, which were ubiquitous across ice age Europe (Gimbutas 2007). However, it is also clear the development of agriculture introduced gender-based tensions and conflicts with the move to permanent settlement, the production of food surpluses and the emergence of incipient notions of control and ownership (Hodder 1990).

Indigenous hunter-gatherer groups in north and west Europe were forced to take on the conflicts as well as opportunities inherent in domesticated food production as they began to adopt it from around 4500 BC. One response to such tensions was to build monuments that enabled them to look back to their origins as well as forwards to memorialise hunter-gatherer conceptions of the community and landscape (Tilley 1994). The passage-grave became a dominant monumental type as agriculture took hold, and this may have been because it functioned very effectively as a place where such groups could negotiate such a transition between ancient traditions and a rapidly changing world.

A ritual cosmology based on the winter solstice enabled Neolithic groups in Britain and Ireland to look backwards as well as forwards, because it represented a critical moment in the seasonal calendar regardless of whether people were practicing hunting and gathering or agriculture, or a combination of the two (Prendergast 2009). The same could be said of birth-based ritual cosmologies. A focus on maternal birth and possibly collective or ancestral rebirth, like a focus on the winter solstice, represented a 'fixed point', a place of origin to refer back and return to.

In the pagan Celtic mythologies of Britain and Ireland, the goddess—or the female aspect of the divine—still had a powerful role. The female Trinity of maiden, mother and crone, well-known in such pantheons, reveals the ongoing need for a divine mother figure as a timeless archetype. It is only in the era of the world religions that the female embodiment of the divine became almost entirely absent. Even so, there is evidence in Christianity for comparable ritual logic—as if the need for a ritual cosmology with a sacred birth at its heart itself has to return as a cultural—and possibly existential trope.

If this is the case, it suggests that the power of 'returning' to some point of ending and beginning is culturally very important. The continued invocation of the image of birth, at Newgrange and in subsequent religious belief, indicates its conservatism—its call to a perceived origin. Birth is perhaps the key motif for representing a return, since it in turn facilitates the very possibility of return itself. It therefore functions as a still point and a point of movement, a moment that both stops time and creates time, that makes death inevitable but also triumphs over it. The mother, the female ancestor, lies behind all cosmologies that elaborate the image of birth. This was especially the case with the great Neolithic monument of Newgrange, which appears to have been designed as the great mother—from whom we are born and to whom we all return—with striking power and elegance.

References

BARRETT, J., (1994), *Fragments From Antiquity*. Oxford, Blackwells

BERGH, S., (1995), *Landscape of the Monuments*. Stockholm

BRADLEY, R., (1998), *The Significance of Monuments*. London, Routledge

BRENNAN, M., (1983), *The Stars and the Stones*. London, Thames and Hudson

COONEY, G. (2000) *Landscapes of Neolithic Ireland*. London, Routledge

EOGAN, G. (1986) *Knowth and the Passage Tombs of* Ireland. London, Thames and Hudson

EOGAN, G. (1991) Prehistoric and early historic culture change at Brugh Na Boinne, *Proc. Roy. Irish Acad.* 91C: 105-32

EVANS-WENTZ, W. (1911) *The Fairy Faith in Celtic Countries*. London H. Frowde

GIMBUTAS. M. (2007) *The Goddesses and Gods of Old Europe: Myths and Cult Images*. Berkeley, University of California Press

HODDER, I. (1990) *The Domestication of Europe*. Oxford: Blackwells

MATTHEWS, C. (1989) *Arthur and the Sovereignty of Britain*. London, Arkana

MEADEN, G. T. (1997), *Stonehenge The Secret of the Solstice*. London, Souvenir Press

MILLER D. (ed.) (1993) *Unwrapping Christmas*. Oxford, Clarendon Press

MORONEY, A.M. (1999),inter Sunsets at Dowth, *Archaeology Ireland* 13: 29-31

O'BRIEN, T. (1988) Winter solstice decoration at Newgrange,*Riocht Na Midhe* Vol 8 No 2: 50-9

O' BRIEN, T., JENNINGS, M., and O' BRIEN, D. (1987) The equinox cycle as recorded at Cairn T Loughcrew, *Riocht Na Midhe* Vol 8 No 1: 3-15

O' KELLY, M. J. (1982) *Newgrange*. London, Thames and Hudson

O' KELLY, M. J. and O' KELLY, C. (1983) The Tumulus of Dowth, County Meath, *Proceedings of the Royal Irish Academy.* 83C: 136-90

PATRICK, J. (1974) Midwinter Sunrise at Newgrange, *Nature* 249: 517-9

PRENDERGAST, F. and RAY, T. (2002) Ancient Astronomical Alignments: Fact or Fiction? *Archaeology Ireland* 16:2

PRENDERGAST, K. (2009) Sacred territories: Astronomy, ritual and the creation of landscape at the passage grave sites of Neolithic Ireland, in Dimitriadis G., (ed.) *Landscape in Mind: Dialogue on Space between Anthropology and Archaeology*, Oxford, British Archaeological Reports

PRENDERGAST, K. (2010) The coming of the light. *Minerva* 21:2

SHERIDAN, A. (1986) Megaliths and megalomania: an account, and interpretation, of the development of passage tombs in Ireland, *Journal of Irish Archaeology* 3:17-30

TILLEY, C. (1994) *A Phenomenology of Landscape. Places, Paths and Monuments.* Oxford/Providence, Berg

Proto-Sardinian Iconography and Ideology of the Cult of the Mother Goddess

Giovanni Ugas

Department of Archaeological Sciences,
University of Cagliari, Sardinia

Abstract. *Sardinia has preserved a huge amount of evidence of the religiosity of its ancient inhabitants from the Neolithic to the Nuragic age, through a remarkable series of sculptural production: reliefs and graffiti on the domus de janas and on menhirs, and also drawings on pottery. Evidence that is no less significant comes from altars, offerings and sacrifices that are well documented both at sacred sites and buildings.*

In this context a place of absolute prominence is occupied by the cult of the Mother Goddess, la Dea Madre. She is portrayed in her wholesomeness, at the start with anthropomorphic features—in the buxom shape of the fertility goddess, then in the stylized forms of the little marble statues in Cycladic style and the Copper Age (Eneolithic) menhirs—until she assumes an aniconic appearance in the Nuragic Bronze Age (second half of the second millennium B.C.) In contrast with the immortal Goddess as an expression of a matriarchal society, her companion—the son who dies—is initially portrayed just by his head, a zoomorphic mask. Then in the late Eneolithic (second half of the third millennium B.C.) he systematically appears in his fully anthropomorphic shape, ithyphallic in the world of the living and inverted in the world of the dead. After the betyl-like aniconism linked to the Nuragic-castle kingdoms, the Mother reappears in anthropomorphic form, with her son, at the dawn of the aristocratic communities of the Early Sardinian Iron Age.

Key words: *Dea Madre, domus de janas; iconography; Mother Goddess; Neolithic; Nuragic.*

The Five Megaliths of the Avebury Cove

G. Terence Meaden

Oxford University Department of Continuing Education (Archaeology) and Kellogg College, 62 Banbury Road, Oxford, OX2 6PN

Abstract. *A review of documents and drawings from the seventeenth to the nineteenth centuries shows that the Avebury Cove, built in the British Late Neolithic, was unique because it was the result of five positioned megaliths. Four were standing stones while the fifth was a huge triangular stone that served as a dry platform or stage. Suggestions are made as to the intended purpose of the Cove stones that as a group align to the midsummer sunrise, and together function to produce a dramatic cult spectacle known as the Marriage of the Gods.*

Key words: *Avebury, Beckhampton, cove, midsummer solstice, Stone 206.*

Introduction

In British archaeology the term 'Cove' is commonly used to denote a megalithic monument of the Neolithic Age comprising three *standing* stones, or sometimes four as at Avebury and Beckhampton in Wessex. There is symmetry of construction that sets a big broad central stone to face a particular desired direction combined with a tall narrow stone standing at each side. At Avebury this three-stone orientation is to the midsummer sunrise while a fourth standing stone—functioning as if it is part of the Cove—is some 25 metres away, almost exactly on the same alignment. At Beckhampton, three kilometres to the west, the orientation of its four-stone cove is to the extreme southerly moonrise.

At Avebury only two megaliths of its Cove have survived the torment of destruction by Christian villagers of the eighteenth and nineteenth centuries (Figs. 1, 2). Formerly there were five. This cove-stone grouping centred a 99 /100-metre diameter ring comprising about 27 stones (Fig. 3).

In 1669 Dr. Walter Charleton reported that the Cove comprised three massive standing stones with an enormous triangular stone lying flat between them (Fig. 4). In addition to any symbolic value that a triangular stone may have been supposed to possess, its presence provided a firm dry platform for activities taking place at that spot.

Dr. Charleton's illustrative sketch of 1669 labelled these four stones as G, G, G and I, the latter being "A triangular stone, of vast magnitude, lying flat on the ground; but (probably) at first imposed on the heads of the other three, in manner of an Architrave." However, the others being of uneven height and far apart, the triangular stone was too small to straddle above them.

Yet the Cove had a fifth stone as well. William Stukeley sketched it in 1722 (Fig. 5) and 1723 and it appears too in a colour painting of 1825 by local artist F. Browne (Fig. 6). William Stukeley labelled this additional megalith Stone F on his master aerial plan of the entire monument at Avebury.

In Fig. 7 Stukeley's sketch of 1723 is modified to include the tall narrow stone that had been destroyed before Stukeley's arrival (Stukeley 1843, Plate 12). It also

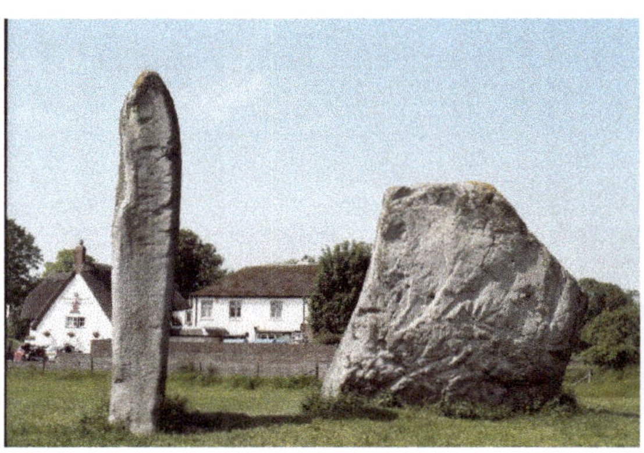

Fig. 1. (Left) Looking towards south-west at the Avebury Cove;

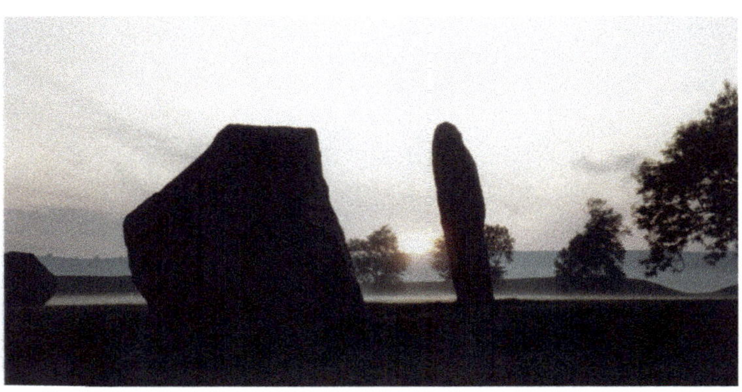

Fig. 2 (right) looking north-east through the gap between two surviving stones to an actual midsummer sunrise

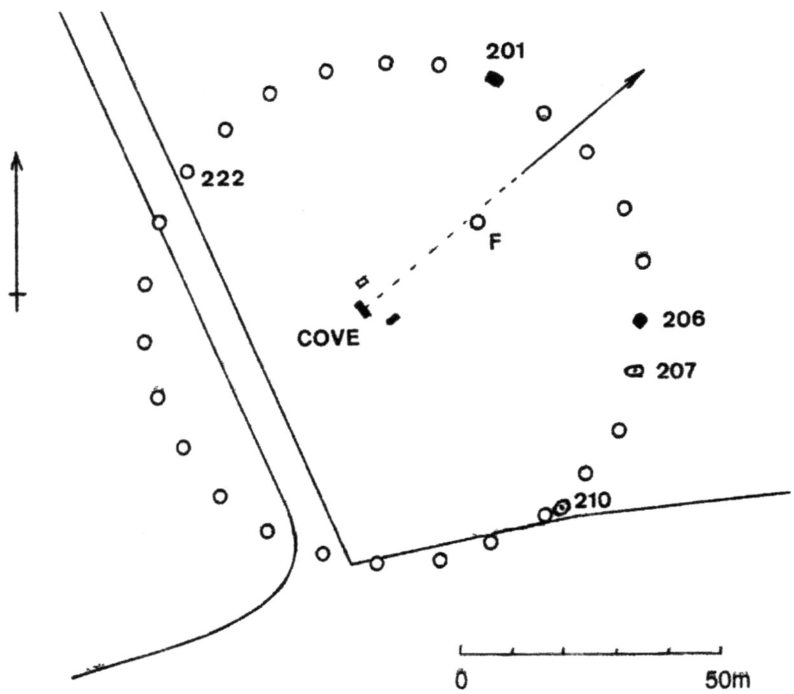

Fig. 3. Plan of the principal stones of the North Circle

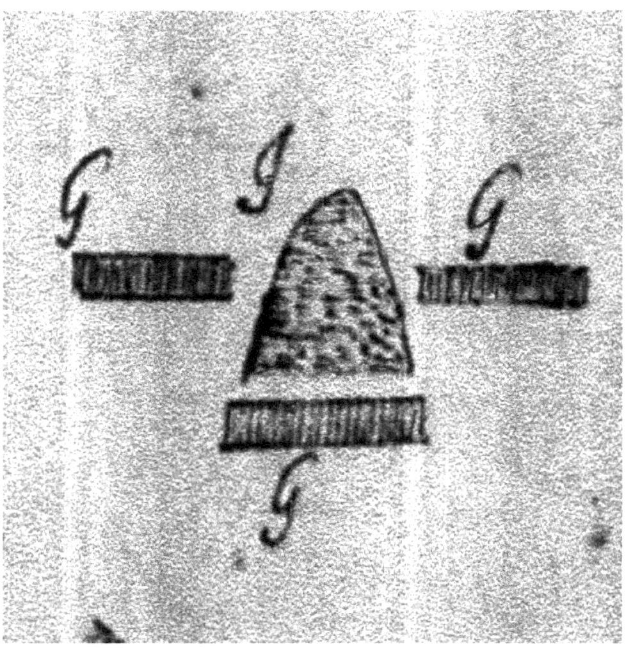

Fig. 4. Part of Dr. Walter Charleton's plan with its schematic indication of the disposition of four closely-grouped cove stones.

illustrates the shadow of the F-stone as it would appear shortly after sunrise on the summer solstice.

Meaning of the monument

Suggestions are now offered as to the meaning of the Avebury Cove stones.

Developing the proposal by Keiller and Piggott (1936) that some of Avebury's stones may have a sexual, hence fertility, connotation, it is possible to propose a realistic interpretation of the meaning of the monument in terms of an agricultural fertility religion, as has been done for Stonehenge. These advances result from applying worldwide knowledge of archaic religions in early agricultural communities to the evidence visible in the planning, disposition and shapes of Avebury's stones.

Archaeologist Isobel Smith (1965), who saw to the publication of the results of

Keiller's excavations expressed a measured view of the shapes of Avebury's stones with these words:

> *There are Type A stones which are much taller than they are wide, with sides more or less parallel and vertical, which are possibly male; whereas Type B stones are much wider, sometimes broader than their height and are possibly female. The latter are squarish, rhomboidal, or trapezium-like, while the best have the shape of a lozenge or diamond balanced on a point.*

On this reckoning the 100-tonne broad middle stone of the cove is pre-eminently feminine, and its two tall narrow

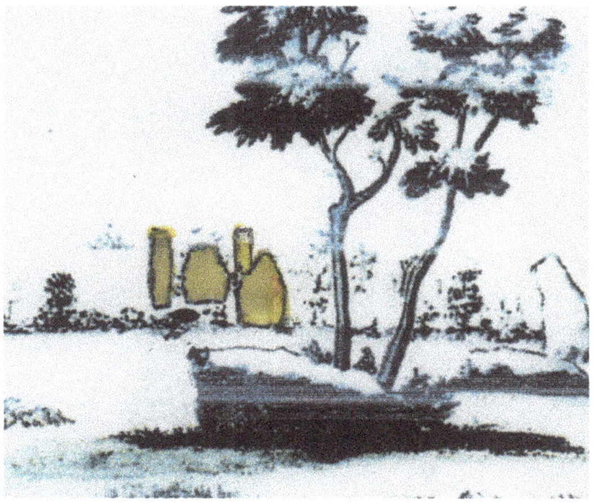

Fig. 5. (left) 1722, a sketch by Stukeley that includes Stone F. The sketch is redrawn to include the missing second tall narrow stone. All four standing stones are here coloured yellow.

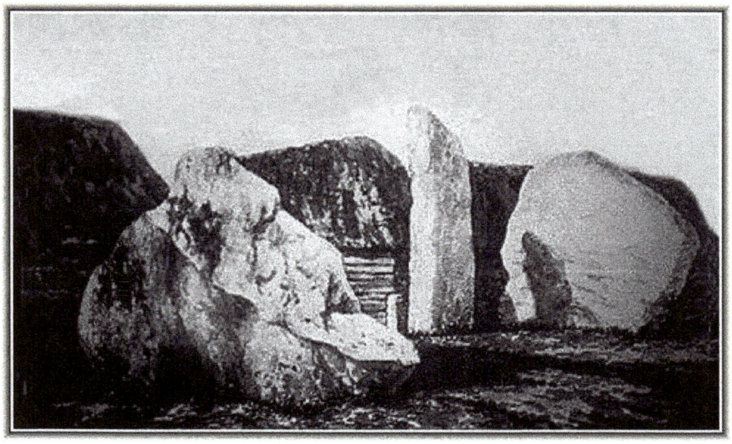

Fig. 6. (right) 1825, painting by F. Browne with Stone F, by then damaged, at the left.

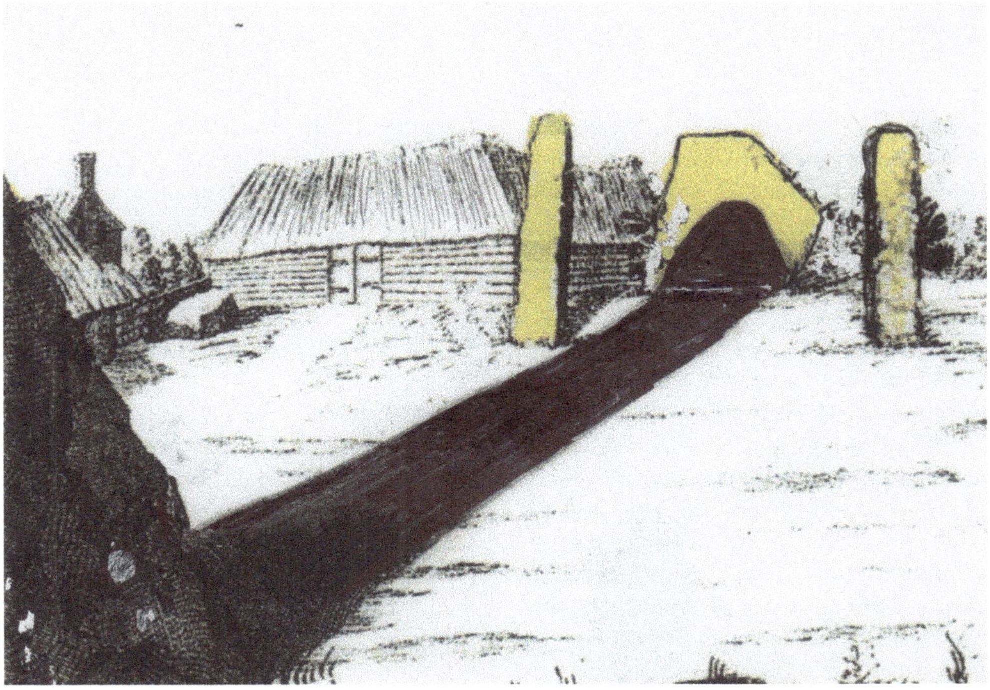

Fig. 7. Sketch by William Stukeley, 1723, redrawn to include the destroyed tall narrow stone so that all four standing megaliths are seen together as at midsummer sunrise.

companions would be phallic and masculine. Stone F is arguably male as well.

This reasoning about the character of the Cove stones is readily reconciled with the universal doctrine that many Neolithic agricultural communities accommodated a belief in primary deities of opposite sex—an Earth Mother and a Sky Father. *This we know because of written sources for numerous historical societies elsewhere, and it can be inferred for prehistoric societies too.* Indeed there are countries where such devotion and worship continue to this day—as with American Native Indians, and tribal Indians of India, Nepal and Borneo among others including Africa.

Crucial is the recognition that the Cove faces midsummer sunrise, on which morning the iconic female cove stone is *bathed in the light of the rising sun until* the shadow of the well-positioned male megalith (Stone F) completes the Midsummer Marriage.

Lastly, it is valid to mention that the broad side of the great Cove Stone facing south-west—namely towards sunset on midwinter's day—has a face carved into it (Fig. 8). This

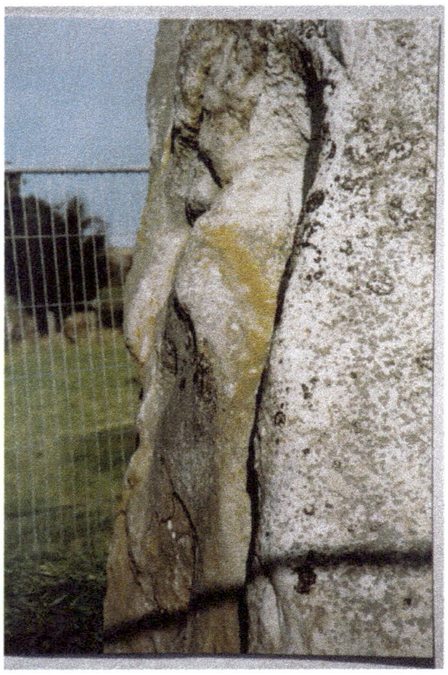

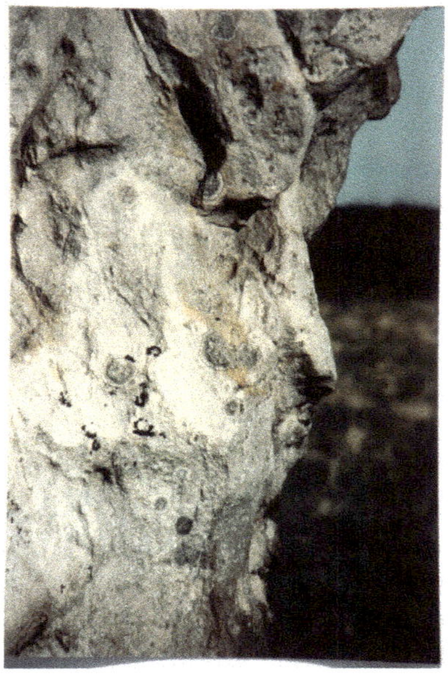

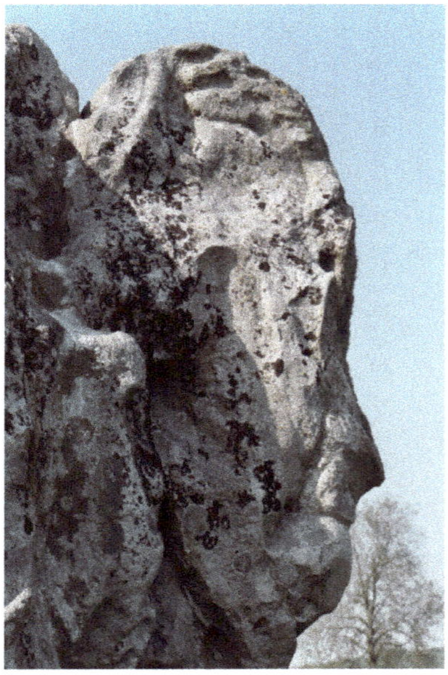

Fig. 8. (Left) The profile of a human head carved into the south-western side of the great Cove stone viewed edge on, looking north-west.

Figs. 9. and 10. (Centre and right) Carved heads on a nearby stone, Stone 206

fine sculpture is a subtle creation because it is visible in profile only over a narrow range of angles. Illumination by the sun in mid-morning facilitates its appearance to the gaze of the patient viewer when, in particular, the bulging eyeball is greatly enhanced. Fig. 9 and Fig. 10 introduce two more Neolithic head carvings present on nearby Stone 206 of the surrounding North Circle.

The Avebury Cove could moreover perform as an actual Sacred Marriage or hierogamy between itself and the sun rising over the north-eastern horizon known as Totterdown. Phonetically, the sound of this place-name differs little from Tar'tar'dun—perhaps derived by combining the widely-known Earth Mother title Tara with her traditional partner, the Sky Father Taran. In a proposed scenario involving the hieros gamos, a feminine Cove functions with the midsummer sun rising over the hill of Tar'tar'dun, to invoke visible matrimony recognizable to a watchful audience as a 'Marriage between Gods' (being a divine union like the Sacred Marriage known to classical Greek poets as between Aeneas and Dido, and Zeus and Hera). Even today, the olden, ill-understood, clarion cry and bugle call 'taran-tara' continues to embolden riding communities in Britain. In the opposite direction south-west of Avebury is the 4-km broad hill range that is Tan Hill, into the western end of which the midwinter sun sets. Again, at Stonehenge to its east-north-east is another Totterdown (i.e. possible Tar'tar'dun) from which the Beltane May sun rises.

Meaden and Howley (2013) describe a similar sunrise scenario invoking the core concept of hierogamy for the Irish recumbent stone circle Drombeg. In 2011 Linda Howley discovered that the eastern of the portal stones bears on its north-eastern side an image of a naked male 'embracing' the stone. The alignment to the circle's recumbent megalith parallels the midsummer sunrise, at which time the portal's phallic shadow encounters the recumbent. This union by shadow had been planned into the monument and can be watched happening on sunny mornings in every midsummer week.

Beckhampton Cove: The Longstones

The megalithic Cove at Beckhampton faces the extreme southerly rising of the moon, namely about 143 degrees east of north. The elevation of Folly Hill to its south-east delays moonrise by 10 minutes. Gradually, the orb of the full moon comes into view, shining on Stone B through the gap between Stone A and Stone C (Fig. 11, Fig. 12). Another critical moment follows when the moonlight is obstructed, and the shadow of Stone C falls upon Stone A. It is the nocturnal consummation between Sky and Earth.

Commentators from Stukeley onwards have remarked on the similarity of shape between coves and the curving forecourts of chambered-barrows, and one wonders whether the closure of the latter coincided with the introduction of the former (Burl 1979, 219). That is to say, were the rituals that were practised in Earlier Neolithic barrow forecourts transferred to Later Neolithic coves? As to their nature, Fowler and Blackwell (1999) comment that "long barrows were dedicated to . . . the Earth Mother . . . monumentalised for all time as sacred places of Mother Earth".

Finally, the last two figures (Fig. 13, Fig. 14) are the author's photographs of two of the Beckhampton Avenue stones found buried near the avenue during the excavations led

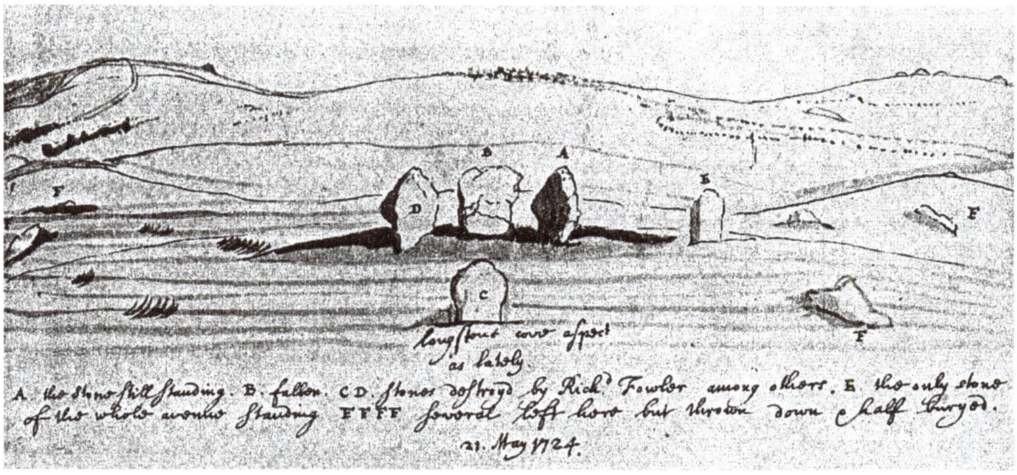

Fig. 11. Stukeley's reconstruction of the cove at Beckhampton (sketch dated 21 May 1724). Stone A of the Cove and Stone E of the Avenue, which survive to this day, are called locally Adam and Eve. At the time of Stukeley's drawing Stone B lay on its back, and several other megaliths F lay fallen. Stones C and D had recently been destroyed.

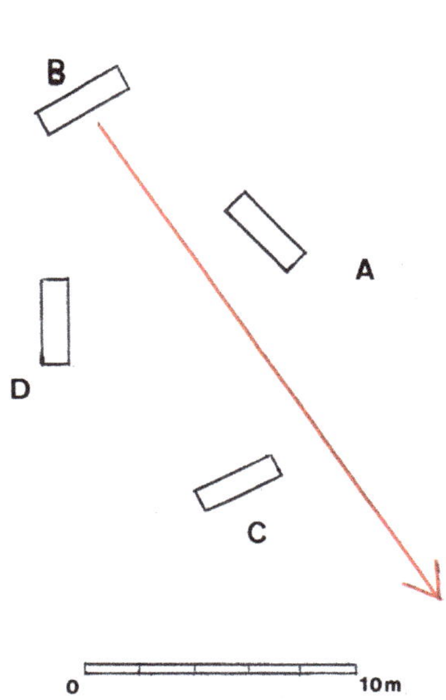

Fig. 12. Beckhampton Cove: Modern plan based on the excavations of 1999 led by Gillings and Pollard. The red line defines the most southerly direction of moonrise over nearby Folly Hill.

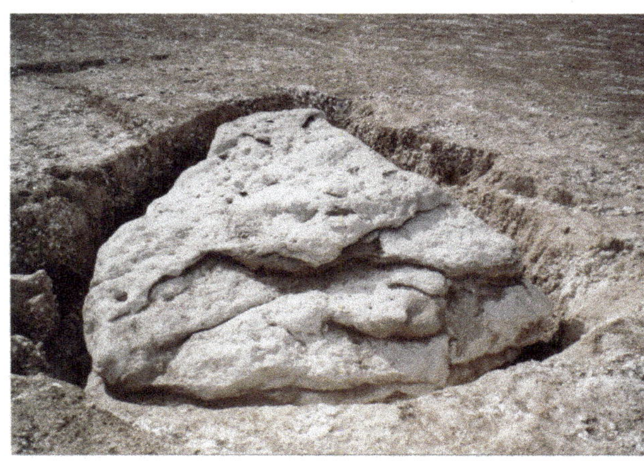

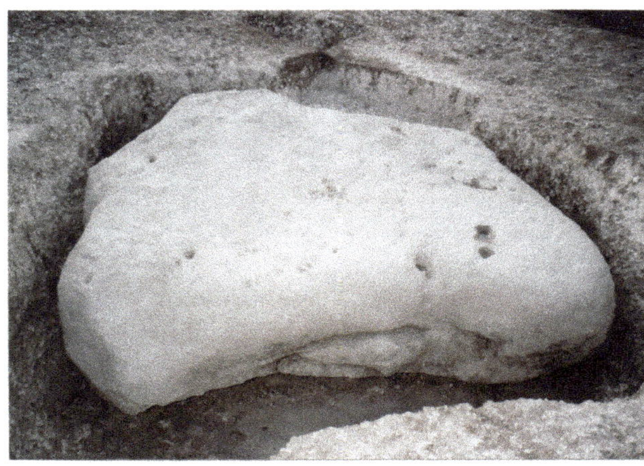

Fig. 13. and Fig. 14. Megaliths of the Beckhampton Avenue uncovered in 1999

by Gillings and Pollard (2004). On symbolic considerations both stones portray feminine symbolism.

Concluding remarks

This study of megaliths at Avebury shows that the three principal stones of Avebury Cove were arranged so that the Cove faced the midsummer sunrise over the hill which is Totterdown, and that the shadow of a fourth standing stone fell upon the middle megalith soon after sunrise. At the Cove a fifth megalith, triangular in shape, lay flat—

probably deliberately so—in order to provide a firm dry platform. At Beckhampton Cove the alignment is to the extreme southerly moonrise.

References

BURL, H.A.W. (1979) *Prehistoric Avebury.* New Haven and London: Yale.

FOWLER P. and BLACKWELL, I. (1999) *The Land of Lettice Sweetapple*. Stroud: Tempus.

GILLINGS, M. and POLLARD, J. (2004). *Avebury*. London, Duckworth.

MEADEN, G.T. (1999). *Secrets of the Avebury stones*, p. 36. Souvenir Press, London. In addition www.megalithic.co.uk/article.php?sid=2146411660 (accessed February 2012) "Was Tara the Earth Mother at Avebury?"

MEADEN, G.T. and HOWLEY, L. (2013). The midsummer sunrise alignment and proposed cult of Sacred Marriage at the Drombeg Stone Circle, County Cork.

SMITH, I. F. (1965). *Windmill Hill and Avebury: Excavations by Alexander Keiller.* Oxford: Clarendon Press.

STUKELEY, W. (1743). *Abury.* London.

Testing Stonehenge Experimentally Using a Replica Full-Size Altar Stone Positioned at the Focus of the Monument

G. Terence Meaden

Oxford University Department of Continuing Education (Archaeology) and Kellogg College, 62 Banbury Road, Oxford, OX2 6PN

***Abstract**: A replica replacement for the Altar Stone was constructed of the same bulk, height and width as the original stone which has for centuries lain fallen, sunken and unavailable. The replica was positioned very close to where the Altar Stone formerly stood, and the sightlines to it were viewed and analysed by direct observation and photography. The conclusion is that the light of the rising sun at midsummer entered the monument four millennia ago (as it still does today) and illuminated the front of the Altar Stone making it sparkle because of its mica light-reflecting platelets. Four minutes later the shadow of the Heel Stone fell upon the Altar Stone. This spectacle could be an action manifestation expressing divine fertility belief—as known from historical and classical sources—of the hieros gamos or the 'Marriage of the Gods'.*

At midwinter sunset the light of the setting sun could only fall upon a very narrow vertical strip of the standing Altar Stone because of blockage by orthostat Stone 55 of the Great Trilithon.

***Key Words**: Altar Stone, Heel Stone, hierogamy, Neolithic, replica, Stonehenge*

Introduction

Of the surviving megaliths at Stonehenge the two most important are the Heel Stone which is external to the monument (Fig. 1) and the ill-named Altar Stone which has sunk into the ground under the weight of heavier stones that felled it when they toppled.

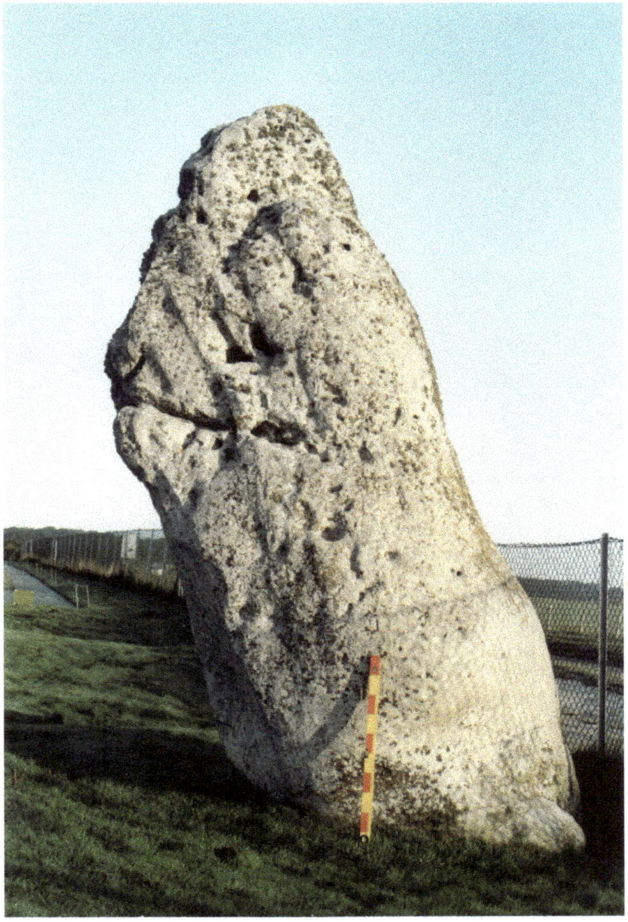

Fig. 1. The leaning Heel Stone, present height 4.7 m. If restored to the vertical, the height is estimated to be 5.2 m

Other key stones are the pairs numbered Stones 55 and 56 and Stones 1 and 30, because these together define the axis of the monument. At the same time the Stonehenge axis is the bisector of the nearest section of the long ditch-and-bank avenue. Both the axis and the avenue correspond with the azimuth of the midsummer sunrise as it was 4500 years ago. Fig. 2 is a map of the monument.

This emphasises that the sun rose over the distant horizon on midsummer morning and shines down the avenue straight into the monument, as it still does, through the trilithon opening between Stones 1 and 30. Sunlight then fell upon the Altar Stone standing vertical and erect as intended when Stonehenge was planned and built. Being a rich mica-filled megalith, it would have sparkled in the sunshine especially if freshly scraped or wetted. Midsummer is the only time of the year when the light of the rising sun can fall squarely on the front of the Altar Stone (cf. Meaden 1997). In total contrast, in the week of midwinter only a narrow beam of sunlight can fall on part of the *back* of the Altar Stone, having passed between Stones 15 and 16 and then between 55 and 56.

The test situation is clarified by **Figs. 3 to 5** that locate the

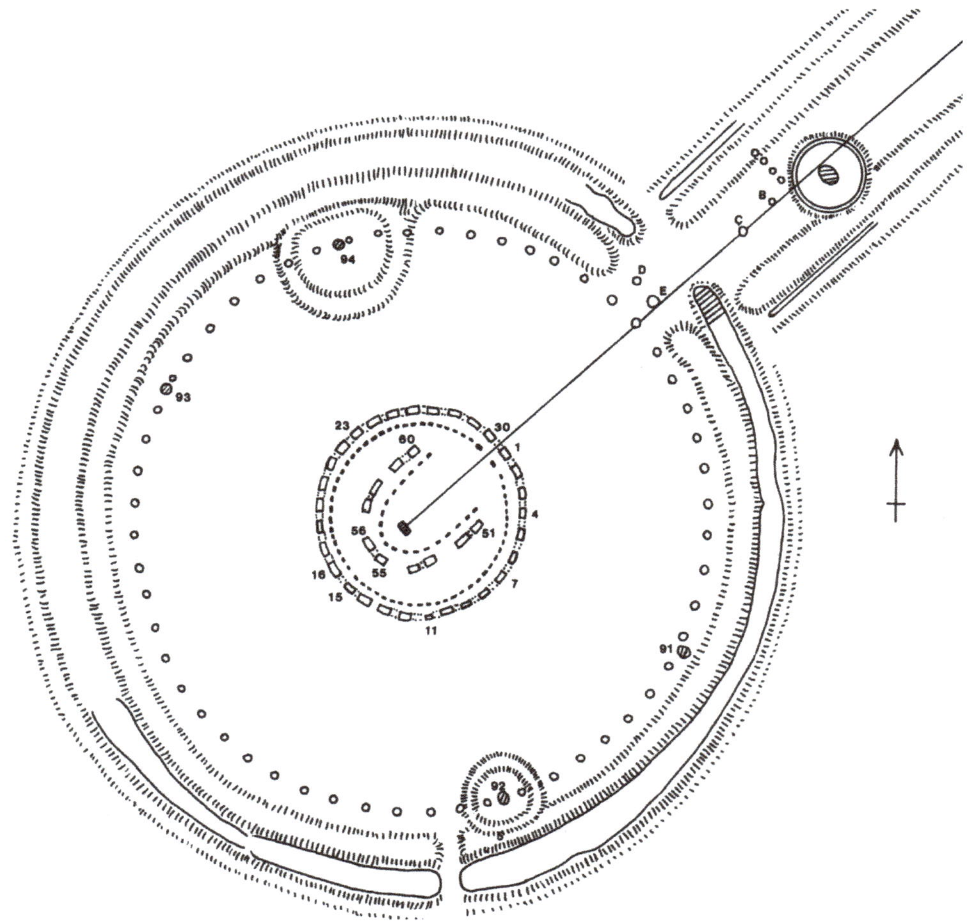

Fig. 2. Map of Stonehenge restored. Note how the paired stones numbered 56/55 and 30/01 define the axis of the monument. This is also a bisector of the Avenue on the final approach to Stonehenge.

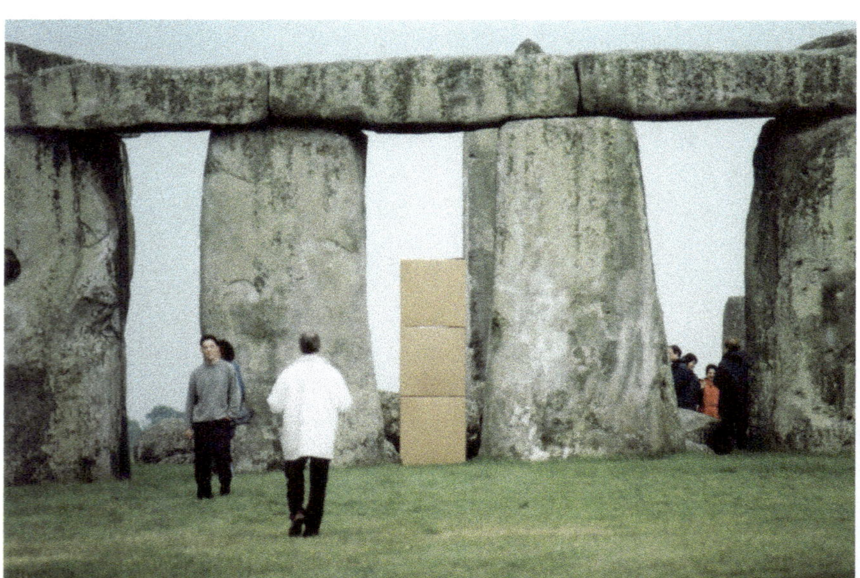

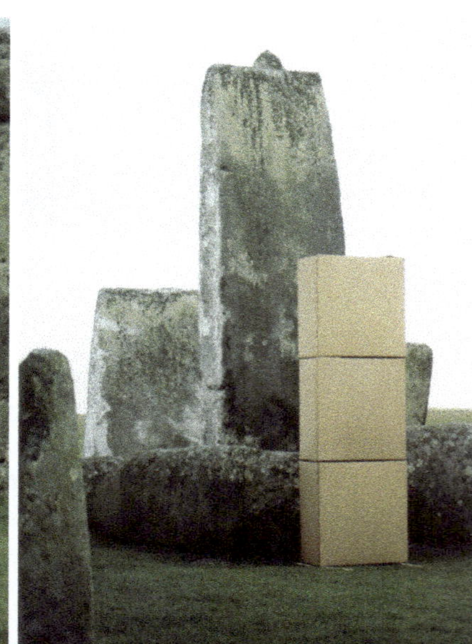

Fig. 3 and Fig. 4. The replica of the Altar Stone is set in position

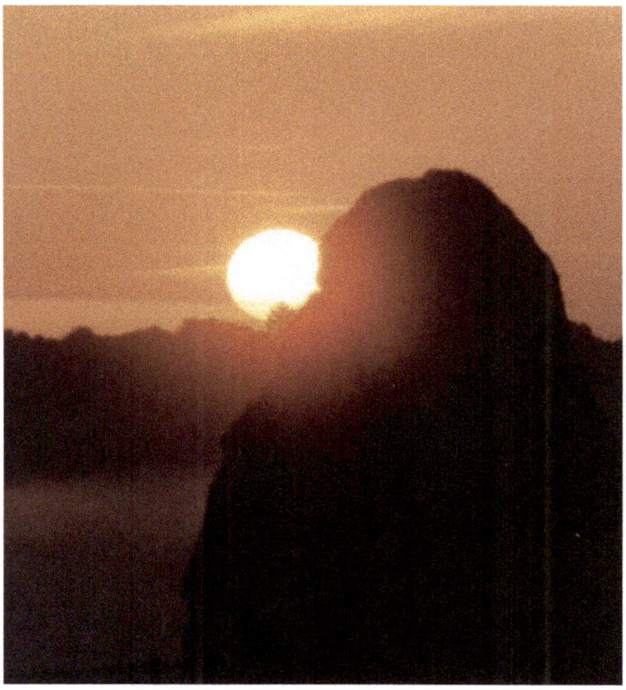

Fig. 5. The midsummer sun begins to rise on the Stonehenge axis before being eclipsed behind the Heel Stone whose shadow enters the monument and reaches the Altar Stone.

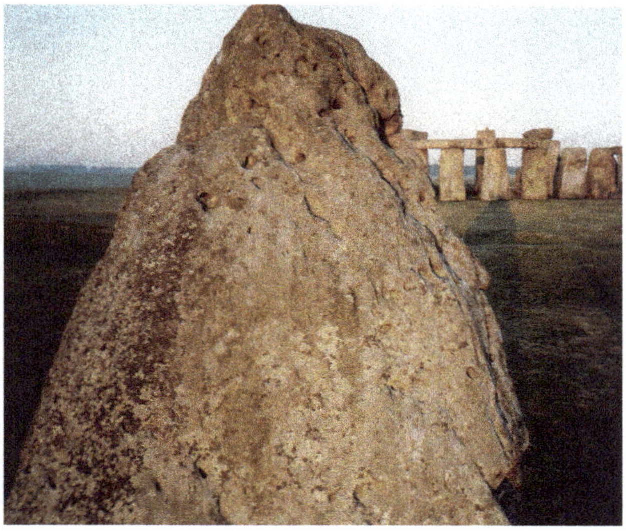

Fig. 6. At the moment of sunrise if the sunshine is weak the Heel Stone shadow is feeble too although it still necessarily enters the medial trilithon arch. A few minutes later, as in this example of 18 June, the sun has got brighter and the shadow has strengthened.

position of the Altar Stone using the replica. The latter was constructed from three cardboard boxes especially made for the occasion by a box manufacturer. The boxes were carried into Stonehenge folded, and then unfolded to stand on the Stonehenge axis as near as possible to where the Altar Stone formerly stood. Each box measured 1m by 1m by 0.5 m, and the total height was 3m which is the same as the unavailable Altar Stone.

Midsummer sunrise

Midsummer photographs were taken on 24 June for which the sunrise azimuth is practically the same as for the day of the solstice which is 21 (or 22) June. The difference in azimuth is a mere tenth of a degree of arc, as it also is for 18 June.

When it was erected, it is likely that the Heel Stone stood vertically (Atkinson 1956, Cleal et al. 1995), not leaning as it does now. When upright it stood half a metre higher. Also 4500 years ago, because of subsequent changes in Earth's ecliptic due to the precession of the equinoxes, the sun rose half a degree further towards north than it does now.

Allowances for these changes, as needed for constructive photography, can be made by the observer squatting in the middle of the monument in front of the Altar Stone instead of standing. The top of the Heel Stone is visible above the horizon as the planners intended (Fig. 5). The midsummer sun then rises a little to the north as further intended, i.e. to the left as viewed from the monument (Fig. 5).

When looking in the reverse direction at midsummer, from a standing position alongside the Heel Stone, one can watch the effect of the Heel Stone's shadow as it crosses the landscape in the foreground of the monument. The darkness of the shadow depends on the strength of the shining sun, but the shadow is always there if the sun is not concealed by cloud. On clear-sky days the shadow is very strong. Even on sunrise days in the week of midsummer if the sunshine is weak the author has seen a faint shadow entering the monument. The next figure illustrates (Fig. 6) and the caption helps to explain.

Just after sunrise on 18 June the shadow had entered the monument between stone numbers 1 and 30, but on this occasion the shadow was too weak to show up photographically. Nonetheless, a little later with brighter sunshine--as shown in Fig. 6—the shadow has strengthened while shifting to the right but is still touching Stone 30. This stone is the right hand (western) orthostat of the trilithon arch through which the sun's rays—followed by the Heel Stone shadow—had earlier passed.

By contrast, on the best days, when the sun shines powerfully because the air is clear at the horizon, the Heel Stone's dark shadow plainly penetrates the monument (cf. the example of Fig. 7 for 24 June).

The ill-named Slaughter Stone plays no role in this, neither positively nor negatively. It lies—because deliberately buried in antiquity—half way between the Heel Stone and the stones of the great circles. The stone had been buried by cutting a deep enough trench of appropriate shape in the chalk. The stone had perhaps been used in a test experiment to enhance the Heel Stone's shadow, but following a change of plan (was this a failed experiment?) it was interred so

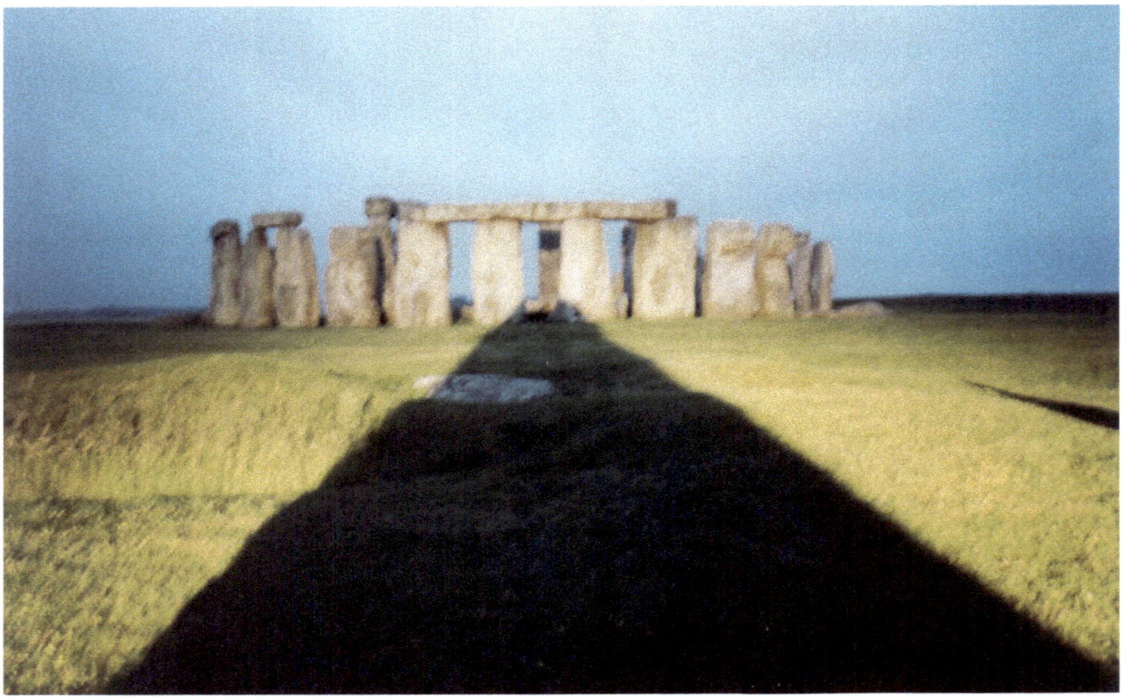

Fig. 7. On a clear-sky morning in midsummer week the bright sun casts a strong shadow through the medial trilithon arch to the centre of the monument.

as to be out of sight and not interfere with the true throw of the Heel Stone's shadow. The Slaughter Stone is visible today only because of the activities of careless excavators in recent centuries.

Midwinter sunset

In midwinter week the sun sets in the opposite direction to that of midsummer sunrise. Observers *today*, if standing on the Stonehenge axis outside Stonehenge *at the north-east*, can watch and photograph the setting sun as demonstrated by Fig. 8. However, contrast this with Fig. 3 where the replica Altar Stone blocks this view—as did the original Altar Stone when it was upright. From this angle the original sky watchers standing on the axis *could never have seen* the sun setting.

Next, consider people watching from the south-western side outside Stonehenge. The line from the sun to the Altar Stone, as sunset approaches, is wholly blocked by the eastern orthostat (number 55) of the Great Trilithon. Then, a few minutes before sunset the blocking shadow of this orthostat (which has been moving to the right, i.e. eastwards) begins to leave the Altar Stone, allowing the downward-moving sun to shine feebly on no more than a narrow strip of the western side of the rear of the Altar Stone.

Carved face

There is a major image purposefully carved into a trilithon orthostat at Stonehenge. This human face is on the west-facing *side* of Stone 54 (Fig. 9, Fig. 10). The image is readily visible when shadows are cast if the sun is shining between 1300 and 1430 British Summer Time or 1200 and 1330 GMT wintertime. The fine image is optimum in the summer months when the sun is high in the sky because this emphasises the eyebrows, nose and lips of the face. The carving had been overlooked until first reported by the author (Meaden 1999).

Conclusions

It is shown that the planners and builders of Stonehenge arranged that in midsummer week the light of the rising sun would fall upon the upright Altar Stone—as it still visibly does 4500 years later when a replica Altar Stone is in position. It is also certain that the shadow of the Heel Stone continues to penetrate the monument and reach the Altar Stone because this has been witnessed. The Heel Stone and its shadow are phallic. The passage through trilithons along the Stonehenge axis to arrive at the Altar Stone appears to be vulvar and vaginal.

In the time of the Neolithic and the Bronze Age when successful fertility of grain, beasts and women was paramount for farming societies throughout the Eurasian Continent, it is probable that one feature of a Mother Earth related religion was the idea of episodic mating between Sky Father and Earth Mother. The concept and practice of ritual hierogamy is well known in later historical times for the classical countries of the Mediterranean and the Near East and Middle East (Frazer 1922, Campbell 1974, Eliade 1958). For both Stonehenge and Avebury it appears that

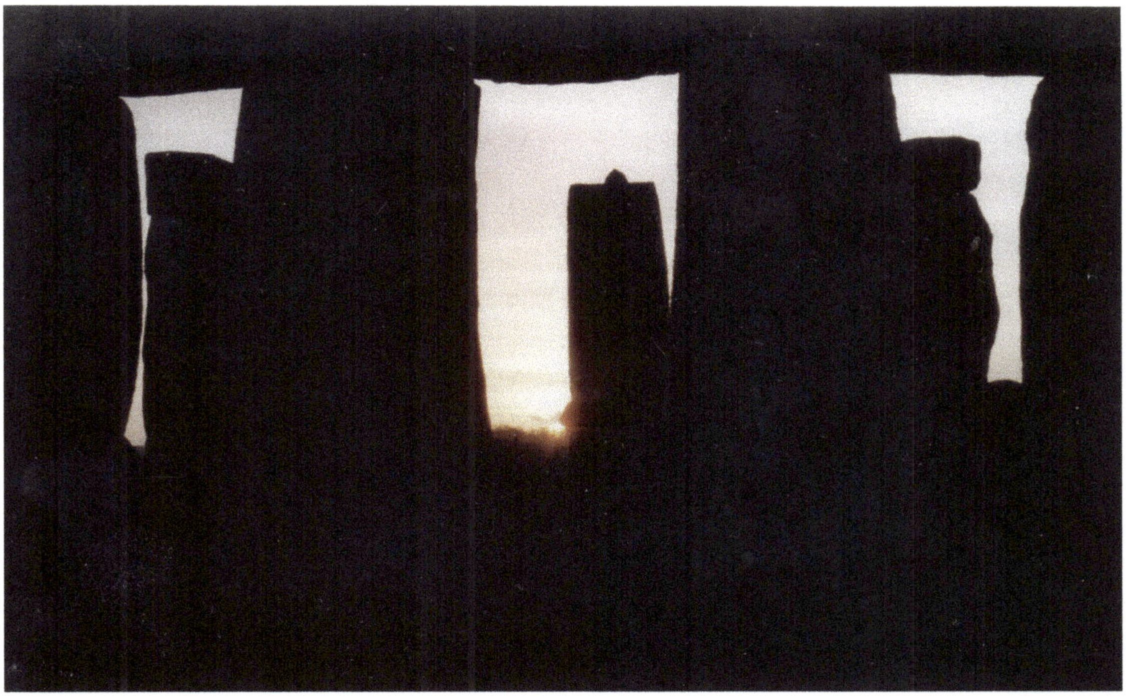

Fig. 8. The setting sun at midwinter when viewed today along the Stonehenge axis from a position alongside the Heel Stone—but in antiquity the partner stone to the surviving orthostat of the great trilithon had not yet fallen, and the Altar Stone was also in the way and still standing.

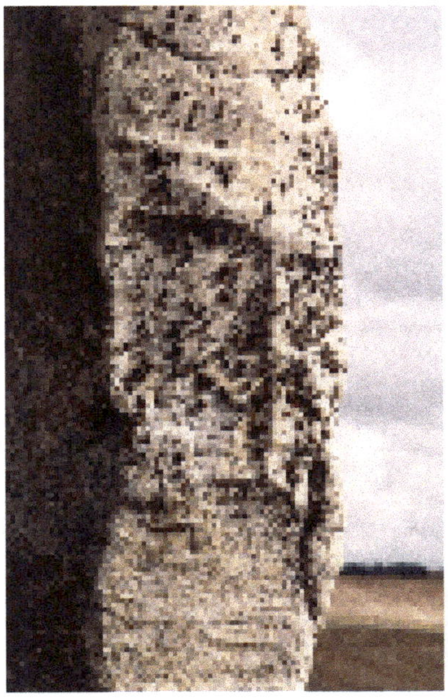

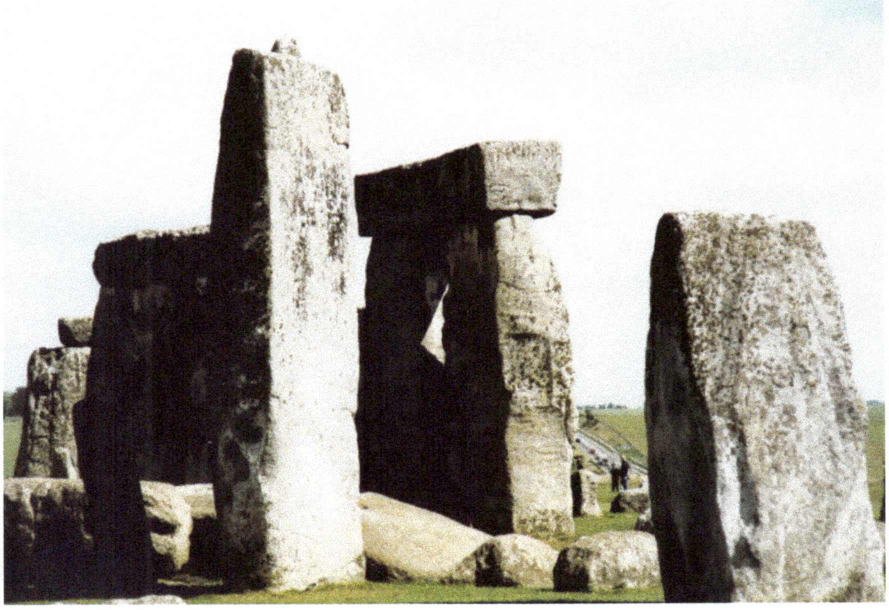

Fig. 9 and Fig. 10. A true carving of a face on the side of Stone 54 at Stonehenge.

a re-enactment of the Sacred Marriage was arranged as a visual spectacle that could be witnessed by many people, and that it took place at the summer solstice.

References

ATKINSON, R. J. C. (1979) *Stonehenge: Archaeology and Interpretation.* London: Pelican (originally 1956 by Hamish Hamilton).

CAMPBELL, J. (1974). *The Mythic Image.* New Jersey: Princeton University Press

CLEAL, R. M. J., WALKER, K.E. and MONTAGUE, R. (1995). *Stonehenge in its Landscape: Twentieth Century Excavations.* London: English Heritage

ELIADE, M. (1958). *Patterns in Comparative Religion.* London: Sheed and Ward.

FRAZER, J. Gordon (1922). *The Golden Bough: A Study in Myth and Religion.* London: MacMillan.

MEADEN, G.T. (1997). *Stonehenge: The Secret of the Solstice*. Souvenir Press, London.

MEADEN, G.T. (1999). *Secrets of the Avebury stones*, p. 136-138. Souvenir Press, London.

Towards the Interpretation of Neolithic Corporeality: Committing 'Mothers' to Earth

Goce Naumov

Institute for History of Art and Archaeology,
University of Skopje,
Republic of Macedonia

Abstract. Neolithic figurines are often the subject of sharp discussion and debate between the world's branches of speculative tradition and more precise academism. The former tends towards seeing a deification of anthropomorphic figurines, while the latter prudently interpret these artefacts as mediators in social relations among individuals and communities. Thus, the first group of researchers exalted the figurines, whereas the second group rendered profane what they interpreted as their evident social function. Both sides use available arguments to confirm their own interpretations and theories, so that whereas traditional thinkers favour and deify the concept of fertility, more cautious thinkers use their own methods to negate, or at least trivialize, it.

This paper aims to take advantage of these opposing views and from them to develop multi-relational interpretations regarding the character of certain Neolithic figurines. In this context figurines from prehistoric settlements in the Republic of Macedonia are used for further observation and comment. Their bodily features and parts thereof would be associated with a "committing to earth" of the previously "deified" figurine concepts, as well as enriching a somewhat pallid hypothesis as to their social engagement. In addition, several ethnographic analogies are applied regarding gender diversity in communal or domestic rites and transformation of social status due to a finalization of fertility functions.

"The Bee Goddess": Another Form of a Great Female Divinity?

The Development of a Symbol in Old European Art and Religion from Neolithic to Mycenaean Times

Angeliki Liveri

Abteilung für Byzantinistik und Neogräzistik
Des Instituts für Altertumskunde der Universität zu Köln
Philosophikum, Albertus-Magnus-Platz, D-50923 Köln

Abstract. *Since the Stone Age, bees have figured among the sacred animals or symbols linked to the worship of a major female divinity like the Mother Goddess, the latter associated with fertility, water, the growth of nature, and regeneration. The oldest representations of bees have been found in various European cultural areas, e.g. in Sesklo, Starčevo, Caranovo, Cucuteni, Sardinia and Corsica. These figures show strange bee-like creatures with uplifted arms or antennae, non-human heads and protrusions from the lower body resembling a tail or stinger. In other cases the figures are more schematic, with upraised arms and bifurcated heads or a ridged body. However, it is difficult to identify these figures with certainty, because they are similar to other signs of a great female divinity through the toad, the Birth-giving Goddess or dancing figures. In Cucuteni an hour-glass shaped "Bee-Goddess" is rendered on a stylized bull's head on a bone plate. In the Bronze Age, in the Minoan culture a few bee figures or bee-like creatures occur on seals, seal-impressions, hieroglyphs or as jewellery. This demonstrates, in my opinion, the holiness of the bee in Minoan Crete, but not the cult of a Bee Goddess. Later, in the Mycenaean culture there appear fewer examples with bee representations, probably under Minoan religious influences, and they are also connected with a main Nature Goddess or other Goddesses of Nature, death, life, regeneration and fertility.*

Finally, one can demonstrate continuity of the holiness of the bee throughout antiquity, from the Neolithic to the Mycenaean Period. Representations of the bee vary from bee-like and abstract stylisations to hybrid or very naturalistic figures.

Key words: *Bee, Bee Goddess, Beekeeping, Great Goddess, Mother Goddess, Mistress of Animals, religion, Neolithic, Bronze Age, Mycenaean Times.*

Zusammenfassung: *Seit der Steinzeit sind Bienen unter den heiligen Tieren oder Symbolen dargestellt. Sie sind mit dem Kult einer grösseren weiblichen Gottheit wie der Muttergöttin/Göttin Erde eng verbunden und symbolisieren Fruchtbarkeit, Wachstum der Natur und Regenerierung. Die ältesten Bienendarstellungen sind in verschiedenen Kulturlandschaften Europas, wie z.B. in Sesklo, Starčevo, Caranovo, Cucuteni, Sardinia und Korsika gefunden. Diese Figuren geben seltsame bienenartige Kreaturen mit erhobenen Händen oder Antennae wieder, Wesen mit unmenschlichen Köpfen, deren unteren Leib wie ein Schwanz oder Stachel hervortritt. In anderen Fällen sind die Figuren schematischer mit erhobenen Händen und Doppelköpfen bzw. gespaltenen Köpfen oder runzeligem Körper dargestellt. Man kann diese Figuren mit Schwierigkeit identifizieren, weil sie sich anderen Zeichen der Grossen weiblichen Gottheit, wie Frosch, Geburt-gebende Göttin oder tanzende Figuren ähneln. Eine "Bienengöttin" in Form einer Wasseruhr ist auf einer Beinplatte aus Cucuteni, die selbst die Form eines Stierkopfes hat, eingeritzt.*

In der Bronze Zeit, in der minoischen Kultur erscheinen manche Bienendarstellungen oder bienenartigen Kreaturen auf Siegeln, Siegelabdrücken, Hieroglyphen oder als Schmuck. Das zeigt meines Erachtens die Heiligkeit der Biene in der minoischen Religion, aber das beweist nicht den Kult einer Bienengöttin. Später, in der mykenischen Kultur kann man weniger Beispielen mit Bienendarstellungen finden; Sie entstanden wahrscheinlich unter minoischen religiösen Einflüssen und sie sind ebenfalls mit einer Naturgöttin bzw. Naturgöttinnen des Todes, des Lebens, der Regenerierung und der Fruchtbarkeit verbunden.

Zusammengefassend, kann man eine Kontinuität in der Heiligkeit der Biene in der Antike, seit der Stein- bis zur mykenischen Zeit feststellen. Die Bienendarstellungen variieren von bienenartigen und abstracten Stilisierungen bis zu Mischwesen oder vollplastischen und naturalistischen Figuren.

Schlüsselwörter: *Biene, Bienengöttin, Bienenzucht, Grosse Mutter, Muttergöttin, Herrin der Tiere, Religion, Neolithikum, Bronze Zeit, Mykenische Zeit.*

Introduction

The magico-religious significance of bees is very old. As has been shown, bees already figured among the sacred animals or symbols in the period between 6500 and 3500 BC (Crane 1983, 19-23, 216-217). Prehistoric man painted, formed in relief, or rendered bees and honey-hunting scenes on rocks and on various cult objects such as vases, plates and stamp-seals. Some believe that the bee was one of the first animal Goddesses, linked to the Great Goddess and

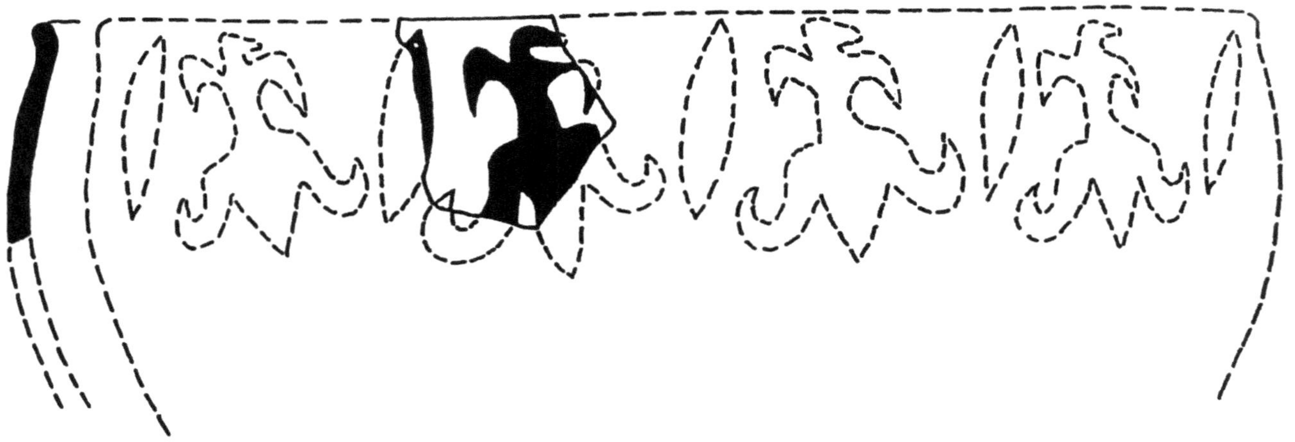

Fig. 1. Frieze of "Bee Goddess" painted on a Proto-Sesklo jar from Otzaki/Thessaly. Partly reconstructed. D.: 6400–6200 BC (drawing after Gimbutas 1991, fig. 142.)

her worship, which was closely associated with fertility, water, the growth of nature and regeneration throughout Old European civilization. It is believed that the Great Goddess is also represented in her epiphany as a bee (Gimbutas 1989, 270; Eadem 1991, 181-185; Golan 2003, 420-422. For the Neolithic Great Goddess see also Hutton 1997; Meskell 1995).

However, scholars always had problems identifying the figures as bees. They either completely overlooked or confused the earliest representations of bees from the Neolithic and Chalcolithic periods of Old Europe with the portrayals of schematized toads or dancing figures. Gimbutas has identified some prehistoric figures as the Bee Goddess: in her opinion all these figures show strange bee-like creatures with uplifted arms or antennae, non-human heads (usually a dot, a stick, or a cylinder) and protrusions from the lower body resembling a tail or stinger. In Old Europe such figures were found at many sites such as Sesklo (Thessaly/Greece), Starčevo (Serbia, Hungary), Karanovo (Vidra/Southern Romania), Cucuteni B (North-West Ukraine), Corsica and Sardinia.

This article presents some of these figures and discusses the development of the bee-motif and its significance in European religion from Neolithic to Mycenaean times. From the Bronze Age to Mycenaean times the article refers mainly to Greek culture.

Neolithic period

According to Marija Gimbutas the earliest European representation of a Goddess in the shape of a bee was found in the Neolithic complexes of Sesklo: On a jar from Otzaki (Fig. 1), a Proto-Sesklo site, there is a figure painted in cherry red on a white background like a frieze, partly reconstructed and dated to about 6400-6200 BC (Gimbutas 1991, 184, fig. 142; Milojčić-von Zumbusch and Milojčić 1971, I, 63; II, pl. before the fig. 1, no 6, pl. E.2; Otto 1985, 45, pl. 3.11, fig.13). Here a creature with wing-like arms and a protruding cone between its parted legs is portrayed. Probably a Bee Goddess is represented, but we cannot be sure because it also shows similarities with dancing scenes or with a birth-giving figure. However, both of these last symbols are also connected with the worship of the great female deity. Dance-figures in particular portray perhaps religious rituals relating to the epiphany of the divinity, or to the regeneration of dying Nature or to other funeral rites (regarding Neolithic dance-figures, see for example Garfinkel 2003; Gimbutas 1989, 241-42, fig. 376-78. On Music: see Melini 2008, 168).

Similar figures with upraised arms, but with differences mainly in the portrayal of their heads appear in other examples too, e.g. on a pottery relief from Truşeşti, classical Cucuteni dated to the second half of fifth millennium BC (Fig. 2), and on an Early Minoan three-sided bead seal of yellow steatite from Kastelli Pediada, southeast of Knossos (Gimbutas 1989, 184, 182, fig. 144,140). The last picture also shows the Goddess with legs parted and a long conical projection which obviously is the lower part of the body of a bee; her arms are upraised, but her head is human. Schematic figures with upraised arms and bifurcated heads have long been known in Starčevo, whereas a headless bee appears in relief on a large vessel from the same site of Kopancs (Gimbutas 1989, 184).

However, it is very difficult to identify these figures with certainty, because, as has been mentioned, they are also similar to a toad or to a birth-giving Goddess, which represent the Great Mother or Mother Earth as well. In addition, the lower part of her body is reminiscent of the letter M, another ideogram of the Great female Divinity. Gimbutas argues that it may identified as a Bee Goddess because of her bee-like attributes, namely a bifurcated insect-type head, legs and a sting (Gimbutas 1989, 184, fig. 143). In her opinion other schematized figures of bees have a characteristically ridged body and are represented quite frequently on linear pottery, e.g. from Holašovice (Fig. 3).

Similar but more abstract representations of bees occur incised in the next examples: (a) on the base of a stamp

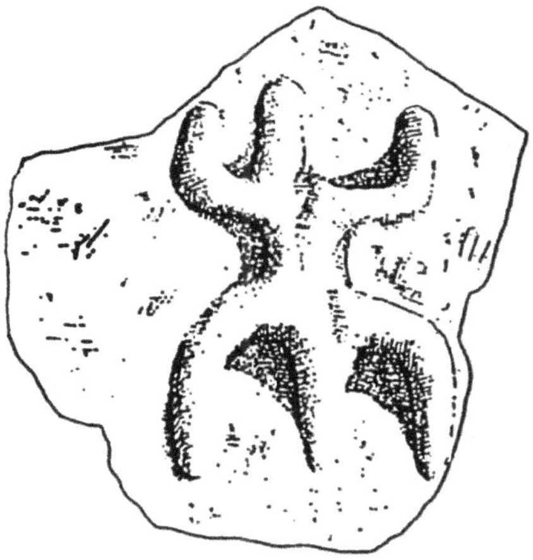

Fig. 2. A "Bee Goddess" in relief on a Potsherd from Truşeşti, Classical Cucuteni. Second half of 5th millennium BC (drawing after Gimbutas 1991, fig. 144.)

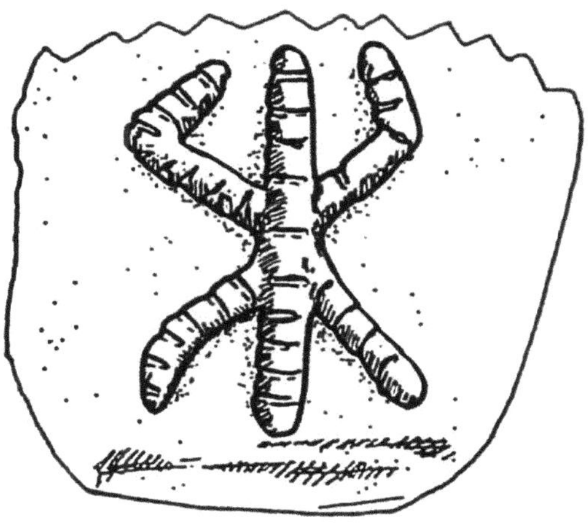

Fig. 3. Schematized "Bee Goddess" on a potsherd from Holašovice (drawing after Gimbutas 1991, fig. 143.)

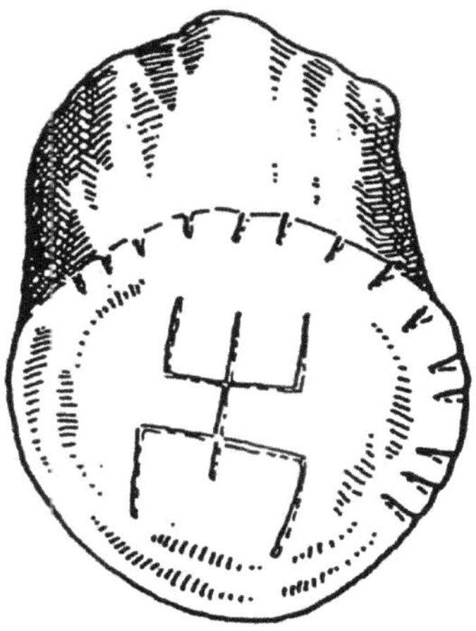

Fig. 4. An incised "Bee Goddess" on a stamp-seal from Karanovo/ Vidra/south eastern Romania. D.: c. mid. 5th mill. BC (drawing after Gimbutas 1989, fig. 422/2b.)

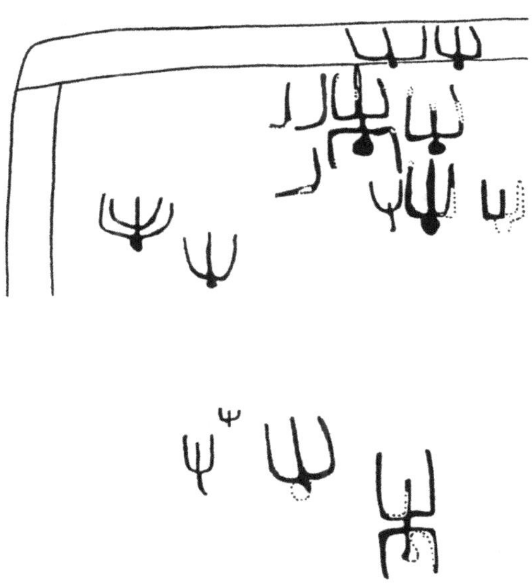

Fig. 5. Anthropomorphized bee engraved on the wall of a hypogeum in Sardinia (bottom right). D.: End 4th mill. BC (drawing after Gimbutas 1989, fig. 423.)

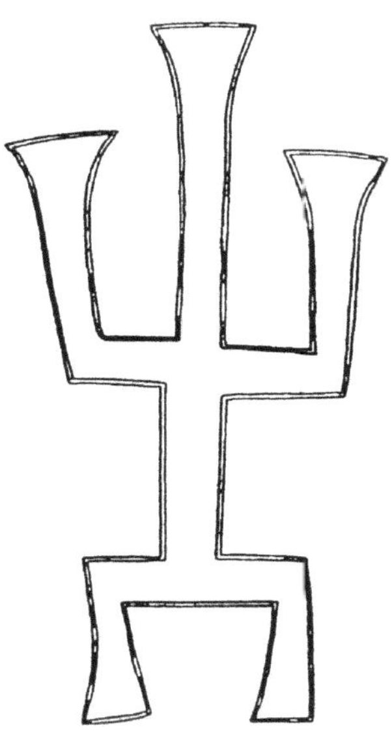

Fig. 6. A schematized birth-giving Goddess engraved on a potsherd from Borsod/north eastern Hungary. Bükk culture. D.: End 6th mill. BC (drawing after Gimbutas 1991, fig. 130.)

seal from Karanovo (Fig. 4), dated to about the middle of the 5th millennium BC, and (b) on the wall of a hypogeum (Tomba dell'Emiciclo a Sas Concas a Oniferi) in Sardinia (Fig. 5), dated to the end of the 4th millennium BC (Atzeni 1975, 22 f., pl. VII/1-3; Atzeni 1981, 69-71, fig. 1, pl. III/3; Atzeni 1982, 102 f., pl. I/2; Gimbutas 1989, 272, fig. 422/2, 423). Similar anthropomorphic representations, painted or incised are found at other sites of Sardinia (e.g. Grotta del Bue Marino di Dorgali, Tomba Branca di Moseddu) and in Corsica (Grotta Scritta di Olmetta di Cup) (Atzeni 1981, 69, fig. 1; Atzeni 1982, 102 f., fig. 8; Atzeni 1989, 160 f., fig. 3/11,12). We can see them also at other European cultural sites, e.g. in Italy (Valcamonica), in Iberia peninsula and in France in connection with the cult of a funerary divinity as the Great Mother (Atzeni 1975, 22 f., nt. 4; Atzeni 1982, 103; Fossati and Jaffe and Simões de Abreu 1991, 13). In these examples a similarity with the birth-giving Goddess (without the middle protrusion, which is a symbol of the baby) is clearer as we can see in the picture that shows a schematized birth-giving Goddess engraved on a potsherd from Borsod in north-eastern Hungary (Fig. 6). It belongs to the Bükk culture and dates back to the end of the 6th millennium BC (Gimbutas 1991, 177, fig. 130). Similar symbols are found in the oldest period of the Vinča culture, 6th or 5th millennium BC. According to Haarmann (1996, 76; Idem 1991, 79, fig. 32a) "if the motifs on the stamps have to be understood as ideograms of the divinity, then the Old European stamp seals are the forerunners of the later Cretan inscribed seals as well". Other scholars (e. g. S. M. M. Winn and J. Hooker) do not agree with this theory. Hooker (1992, 100-112) supported the auchthonous character of Linear A script in Crete and its parallel use with Hieroglyphics. For him, contrary to Haarmann (1989, 251-75), a continuity in the 'Writing' from Old Europe to Ancient Crete is not plausible.

A punctuated silhouette of a Bee Goddess appears on a stylized bull's head of bone (Fig. 7) from the Cucuteni site of Bilcze Zlote, dated to late Neolithic times (3700-3500 BC; H. 17cm). (Crane 1983, 217, fig. 260; Gimbutas 1989, 270, fig. 420; Eadem 1991, 246, fig. 7-39, 178; Golan 2003, 131, fig. 124/2). Here the bee, an attribute of the Mother Goddess, is associated with the bull or with the horns of consecrations (compare with Catal Hüyük). In Cucuteni the Mother Goddess is portrayed as a bee or wasp-wasted figure, a so-called hourglass figure. Her body consists of two triangles (an hourglass) joined in the lower part with a third triangle, a pubic triangle; her upraised arms are bifurcated and her head is a line of many dots. This picture is also very important and unique in the European art of this period.

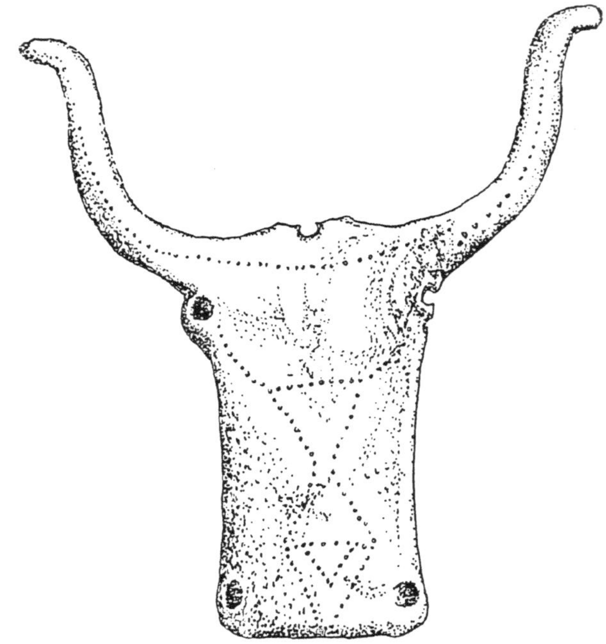

Fig. 7. A punctuated sihouette of a "Bee Goddess" on a stylized bull's head of bone from Bilcze Zlote/northwestern Ukraine. Late Cucuteni, 3700-3500 BC. H 17cm. (Drawing after Crane 1983, fig. 260.)

Hourglass figures are represented in many objects and sites and are connected with female divinities of water, birth, death and regeneration. (Gimbutas 1989, fig. 373/1, 2; Eadem 1991, 7-28/2, 132, fig. 92, 110; Golan 2003, 41, fig. 36/3). A very strange hybrid hourglass Goddess (Fig. 8) is painted on a cave wall in southern Spain (Los Organos, Despeňaperros, Jaen). She has the body of a woman, the feet and arms of a bird and a triangular head that has three horizontal lines emanating from the cheeks, protruding circular eyes and the antennae of an insect (perhaps of a bee). This figure is dated probably to 4th millennium BC and is identified with the Goddess of Death and Regeneration, who is at once a woman, a bird of prey and a bee (Gimbutas 1989, 240, fig. 376/2; Eadem 1991, 246, 248, fig. 7-40).

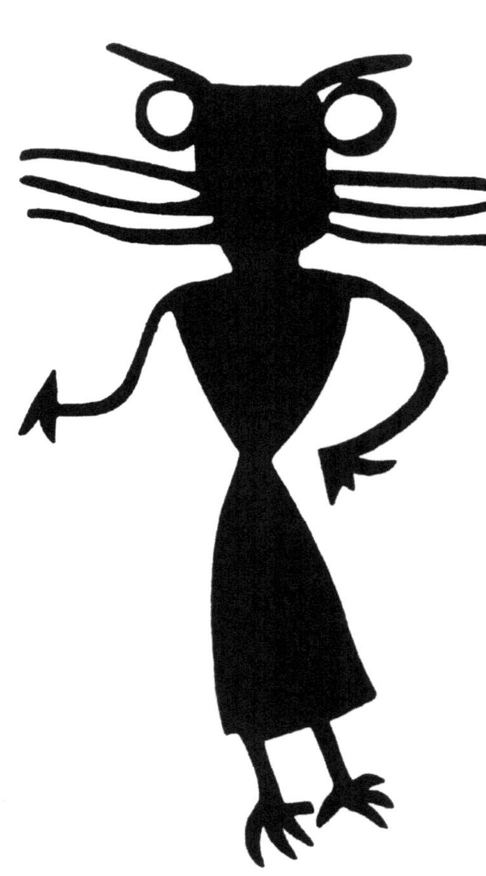

Fig. 8. A Hybrid hourglass-shaped Goddess painted on a cave wall Los Organos, Despeňaperros/South Spain. D.: 4th mill. BC (drawing after Gimbutas 1989, fig. 376/2.)

Concerning the bull-symbol: it was found not only in the Cucuteni cultural area (in the shrine of Sabatinovka in Moldavia), but also in other European Neolithic sites such as in the Vinča area, in the Ozieri culture (hypogea in Malta and Sardinia) and in Cyclades (5th -4th millennium BC) (Gimbutas 1989, 269, fig. 416-17; Eadem 1991, 286-92, fig. 7-108, 112, 113: hypogea; fig. 7-36, 7-37: anthropomorphic vases from Early Cycladic and Classical Cucuteni cultures. For representations of bulls from Tarxien Temple on Malta see also Trump 2005, 120-127, fig. p. 124; Pace 2006, 31-32; Sharon 2006, 22-23). The bull was the symbol of virility, of the Great male deity, of the Great God, the consort of the Great female Goddess. It is in this connection that the bull symbol is probably to be understood. (On the symbolism of the bull see: Catalogue Exhibitions 2006, 41-44; Gimbutas 1989, 265; Mahlstedt 2004, 59-60, 94-96, fig. 10).

The cult of the bull or of the bull's head with its horns

(horns of consecration) is very old. Representations of bulls with horses or lions decorate Palaeolithic rock-paintings, e.g. in Peche-Merle, Lascaux and Parpalló. However, according to Jacques Cauvin the pair bull-female figure is the most important of these pair-symbols. In his opinion, "the symbolic figures of the woman and the bull appear for the first time in Syria-Palastina about 9500 BC. These figures dominate throughout Neolithic and Bronze times in the East and in pre-Greek religions of the eastern Mediterranean" (Cauvin 2004, 25). Çatal Höyük "shrines" or houses of the 7th millennium BC are dominated by woman and bull representations (idols, wall-paintings, reliefs). (Cauvin 2004, 25-30, fig. 9/2, 10; Cutting 2007, 126-128, 130-134; Dietrich 1974, 99-101; Hodder 2007, 124 f. ; Mahlstedt 2004, 94-96, fig. 10/5; Mellaart 1967, 95 ff., 181 f., pl. 37, 38, fig. 45). In other Neolithic sites of Anatolia bull representations have been found as well.

As has remarked by Dorothy Cameron (1981, 4-5), the Neolithic people associated the bucranium with the female uterus and fallopian tubes, and because of this extraordinary likeness the head and the horns of the bull were symbols of fertility and regeneration. The cult of the bull and of the horns of consecration existed in European Neolithic cultures and still survived albeit transformed, in Cyprus, Minoan Crete and in classical Greece (Catalogue Exhibitions 2003; Catalogue Exhibitions 2006, 43 f., no 25-32, 41, 71, 84, 92; Dietrich 1974, 100, 114-117; Gimbutas 1989, 265-71; Haarmann 1991, 86-87; Matz 1961-62, 215-223; Moss 2005, 196 f., 198 f., 200, 210-213, fig. A1.1-4; Nilsson 1927/r.1950, 165-193, 357-368; Rutkowski 1981, 75-90, 116, with many figures and pl. 5,2, 6,2, 8). Dietrich (1974, 115) agrees with Picard and sees "in the famous Cretan bull games not a "secular sport", but a religious rite which formed a part of a *drame chthonien*". Others see the bull as a sacrificial victim, but also sacred. It must also be mentioned that some scholars such Nano Marinatos (followed Friedrich Matz) have supported the opinion relating to the Bull Hunting and Bull Leaping scenes in Minoan Crete that "the bull is not a god nor was worshipped by the Minoans. Rather it was a powerful animal which could be conquered by man" (Marinatos 1993, 73-74; Eadem 1994, 89-93; Eadem 1999, 481). Moss (2005, 196, 198) believes that in Minoan Crete "the horns of consecrations may have originated in Anatolia, but their use as a religious symbol seems to be connected to the iconography of the Egyptian cow-headed Goddess Hathor, denoting renewal and protection of the solar god". However, she sees the origins of the bovine head in Egypt, but also in connection with the cult of Hathor. (On the origins and symbolic of the horns of consecrations compare also Rutkowski 1981, 87-90. See also representations of bucrania on seal-impressions e.g. CMS XII, 65c; CMS XII, 49; Papapostolou 1977, no 22, p. 64-66, pl. 2b, 32b, 33b, c. (from Chania), p. 119 (from elsewhere, *prosopeio* of a bull?).

In Greek Religion (from Minoan, Mycenaean to Classical and Hellenistic times) the belief survived that the bull was a part of the Mother Goddess, her son. His cult was probably connected with the natural cycle of life and death as later myths on Divine Male Childs such as of a youthful Zeus, Dionysus, Hyacinthus and Erechtheas/ Erichthonius show. Zeus was identified with the Cretan Velchanos and pictured on Cretan coins with a bird and the bull. Dionysus as companion of a Great Goddess of Nature appears many times as tauromorphic, and symbolized the growth and decay of vegetation; the name of his mother, Semele signified Earth. This is known in the Mycenaean cult of Hyacinthus at Amyclae/Peloponnese, from Minoan origins, and in the festival of Artemis Hyakinthotrophos in Cnidos as well (Dietrich 1974, 116 f., 172-175). It seems that the myth of Europa and the bull was born from this association of a Mother Goddess and her companion in the shape of a bull; in addition the association of Heaven and Sun or Moon already saw the essential nature of the bull-Zeus in this legend (Dietrich 1974, 119, nt. 303. Compare also the Minotaur myth and later Artemis Tauropolos. For representations of the Minotaur see in Kakridis 1986, 47 f., fig. 25, fig. in p. 258, p. 269 f., fig. 146; LIMC VI/1,2 1992, 571-572, 574-581 (Minos, Minotauros), fig.17-24 in p. 312-315, 35 and fig. in p. 316-325, fig. 3-72 ; Schlager 1989, 224-237, fig. 1-10. For Cypriot "bull-men" see Karageorghis 1971, 261-63, fig. 1-7. On the Artemis Tauropolos: LIMC II/1, 2 (1984) 624 f., no. 700-705).

The Minoan belief that after the sacrifice of the bull a new life would emerge is very interesting. The expression of the new life would be a bee or a butterfly symbolically emerging from the cadaver.

Dietrich (1974, 119-22) and other scholars (Crane 1979, 453 f.; Davaras 1984, 85 f., fig. 4; Mellaart 1967, 186 f., pl. 41-42, fig. 46;) see influences from Anatolia (Çatal-Höyuk) later in the Minoan religion, e.g. a relationship between the symbolism of the life of the bee (life and death, renewal symbol), the Hittite god Telepinu and the Cretan Zeus: Telepinu disappears every year and he was found by a bee again sent from the Great Goddess. The Cretan Zeus, a vegetation god, died and was reborn every year as well. We can also compare this with the myth of the nurturing of the infant Zeus by bees in a Cretan cave. Later, Dionysos-Zagreus is resurrected in the form of a bee. Bees also symbolize the souls and their resurrection (Davaras 1984, 85 f., fig. 4, 90, nt. 5; Dietrich 1974, 119-122; Picard 1948, 230).

Bronze and Iron age (Minoan and Mycenaean Culture)

In the Greek region, after Sesklo, the bee motif appears much later, i.e. during the Bronze Age in the second millennium BC in the Minoan culture. In Minoan religion some insects, among them bees, were very important and sacred as evidenced by their existence in hieroglyphic script, on seals or seal-impressions and as funeral or god offerings as well. In Crete the bee was probably an attribute of the Mother Goddess, an expression of her epiphany, but less schematized than in Neolithic times. At this point is important to mention the question whether one accepts one central or main Great Divinity, a Nature Goddess like Eileithyia *(Ειλείθυια)* or many deities (gods and goddesses) in Minoan Crete or later in the Mycenaean Religion (see Dietrich 1974, 128-190; Moss 2005, 1-f. with all theories; Rutkowski 1981, 107-121. Compare Catalogue Exhibitions 2008, 190-210; Thomas-Wedde 2001, 3 f., nt. 3 with bibliography). The scholars are not agreed. Godart (2001, 463-66) for instance argues for a "Terre Mère" with the name ma-ka for the Mycenaeans and later Demeter. The polytheism of the Aegean Bronze Age is uncertain and cannot be indicated, but "this of the Dark Age is a well-established fact." However, the bee must be connected with a deity or deities of Nature, vegetation, fertility, death, life and regeneration.

Representation of bees

The bee is sometimes depicted on one face of Minoan seals, seal-impressions and gemstones, while on the other face there was a bull or dog, related to the Goddess as well; these date from Early Minoan (EM) until Late Minoan (LM) times. Because the design of these insects on the Minoan seals is so crudely executed, it is difficult to identify them with certainty. There are representations of bees in hieroglyphs which help to identify similar insects on the seals (*History of the Greek Nation* 1971, 149, no 86. Compare CMS II, no 80; Kitchell 1981, fig. 1, 4, 5). It is especially difficult to distinguish the spider from the bee or wasp: the number of legs poses difficulties, as do the proportions of the bodies. For instance, Gimbutas has interpreted as bees the figures on two Minoan seals, whereas other scholars interpret them as spiders (CMS II, no 99; CMS II/5, no 313; Gimbutas 1989, 272, fig. 425; Yule 1981, 134, pl. 9A/7, 8). Davaras (1984, 83) is also unsure of the identification of some insects as wasps or bees on seal-impressions from Phaistos (CMS II/5, no 314, 315-6), from elsewhere (CMS VII, no 70, 86 (?), or in three-sided prisms (CMS VII, no15b (?) and XII, no 42b) and in talismans in combination with horns of consecration (CMS IX, no 97). The same scholar sees a bee in a seal-impression of Lerna, whereas M. Heath Wienke identifies this figure as a spider (CMS V.1, no 58, L 4417; Davaras 1984, 83, nt. 4). Probably some insects have also been identified as flies instead of bees (CMS II/1, no 379). According to Renfrew (1972, 287) these seals were used to stamp consignments of honey, but this is uncertain. Also others think that probably the Early Helladic seals from Lerna are the oldest testimony of trade and among them one can distinguish only between bee- and jar representations. Nevertheless, these last motifs appear on Minoan seals as well. Doumas also recognised on jars from Akrotiri in Santorini a primitive netting shape of a beehive (Doumas 1983, 119, fig. 19; Richards-Mantzoulinou 1979, 79). Cretan seals or seal-impressions present a few various types of bees. The earliest shows two schematic bees arranged tête-bêche in axial symmetry, carved on an ivory ring, dated to the EMII and the MMIA period (CMSII/1, no 159; Yule 1981, 134-35, no 9, pl. 9B/9e). The other bee motifs are more or less schematized, sometimes with heads similar to a bird or to a helix, and they date to MMII or to MMIIB (CMS XII, no 109a; CMS II/5, no 315, 314; Yule 1981, 134, no 10, 11, 12). The representation of three bees en-face with astral motives on a seal-impression from Knossos in the Archaeological Museum of Herakleion is very interesting (CMS II/8:1, no 149). Also in the same museum there is a bee pictured with a dot among its legs and with other symbols and hieroglyphics on a three-sided prism-bead or nodulus dated to MM II period (CMS II/8:1, no 80). It is comparable to bees of Malia (Fig. 9, see below). More naturalistic is the bee on a lentoid of black jasper in the British Museum (CMS VII, no 70. Bees are represented probably in nos 15 and 86 as well). Papapostolou (1977, 63 f., no 20, 21, pl. 2c,d, fig. 30b, 31c, 32a, 33a) also identifies some insects on seals-impressions from Chania and he believes that probably one of them (no 21) represents a bee. Mirié (1979, 78) supposes the existence of a fertility Goddess on Minoan seal-impressions with a body of a bee and the head of a woman with the central characteristic of being able to change.

In addition, scholars have recognised among the Minoan hieroglyphs two (numbers 50 and 51) which represent a jar with tiny specks and have identified them as signs for grain or honey jars or for pollen jars (Evans 1909, 201, 212; Kitchell 1981, 13, fig. 3).

The theory that female figures representing priestesses or worshippers on Minoan gold ring-seals or generally Minoan gemstones are insectomorphic, is unacceptable. Supposed support for the insectomorphic theory comes from, e.g., the gold ring of Isopata (near Knossos) in the Archaeological Museum of Herakleion (Fig. 10), dated to LMII period (1600-1400 BC). (Cain 2001, 33-46, fig. 1, 2; CMS II/3, no 51; Liveri 2008, 9 f., col. pl. II/1; Marinatos 1993, 163, fig. 149; Niemeier 1989, 165, 177-78; Sakellarakis 1994, p. 59, no 424, fig. p. 61 above; Sapouna-Sakellaraki 1994, 325, no 81, fig. 81). Gimbutas (1974, 66; Eadem 1991, 185, fig. 146) and followed her Haarmann (1991, 88, fig. 38 right; Idem 1995, 61-62) believe that on this ring the heads and hands of the human figures are depicted like those of an insect, but we cannot be sure of what kind, nor that the female figures wore insectomorphic masks. Rutkowski (1981, 104, fig. 30/4,12, 31/5-13) supports also the use of the horns- or animal-masks in general crowned by a "snake frame" in the Minoan-Mycenaean culture. Papaefthymiou-Papanthimou (1997, 98 f.) agrees with Hooker (1992, 101) and also doubts the use of masks by the Minoans. Sapouna-Sakellaraki thinks that the human figures of this ring have heads of a bird. Comparable is the gold ring-seal from Archanes, also in the Archaeological Museum of Herakleion (Fig. 11), dated to LM I (1600-1500 BC) (Liveri 2008, 9-f., col. pl. II/2; Sakellarakis 1994, 59, no 989, fig. p. 61 mid.; Sapouna-Sakellaraki 1994, no 80, fig. 80). It shows the same rendering of the heads of the figures as well. The same debate also surrounded the High Priestess-Goddess represented in her epiphany (Fig. 12) on an onyx gem of Knossos (Archaeological Museum of Herakleion, LM II, *c.*

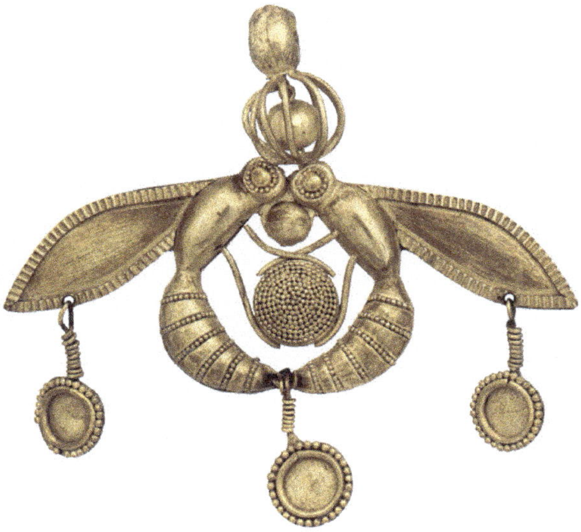

Fig. 9. Gold pendant from Malia/Crete. Archaeological Museum of Herakleion. D.: MMIII, c. 1700 BC. (Photo: Archaeological Museum of Herakleion © HMC-ARF)

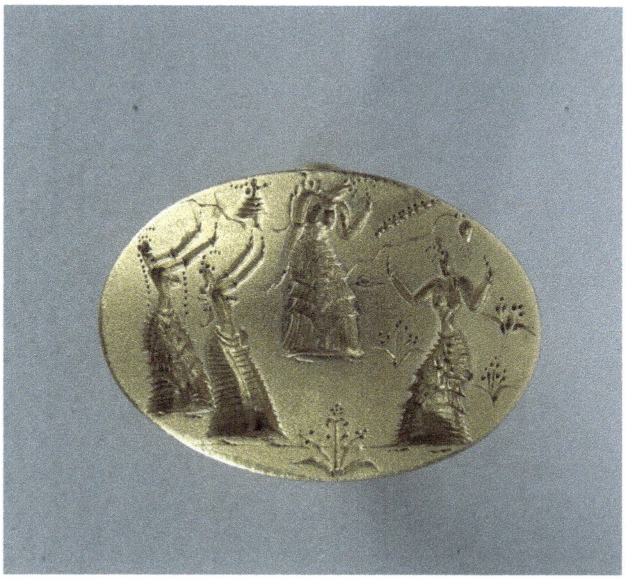

Fig. 10. The gold ring-seal from Isopata. Archaeological Museum of Herakleion. D.: LMII, 1600-1400 BC. (Photo Archaeological Museum of Herakleion © HMC-ARF)

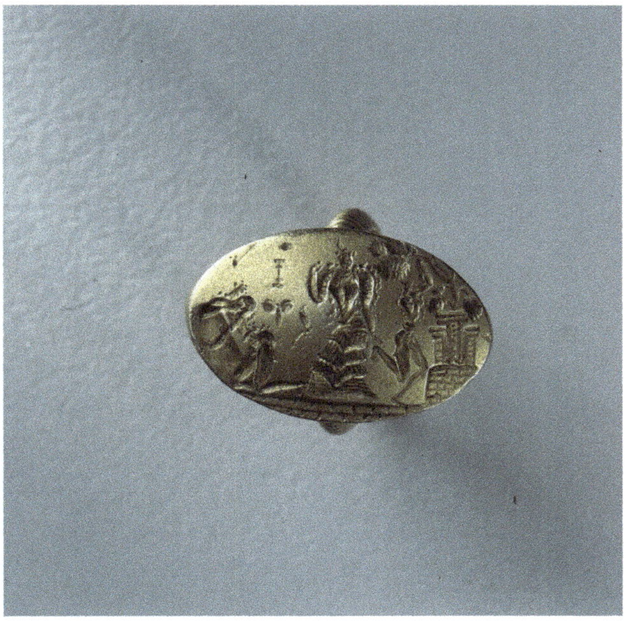

Fig. 11. The gold ring-seal from Archanes. Archaeological Museum of Herakleion. D.: LMI, 1600-1500 BC. (Photo: Archaeological Museum of Herakleion © HMC-ARF)

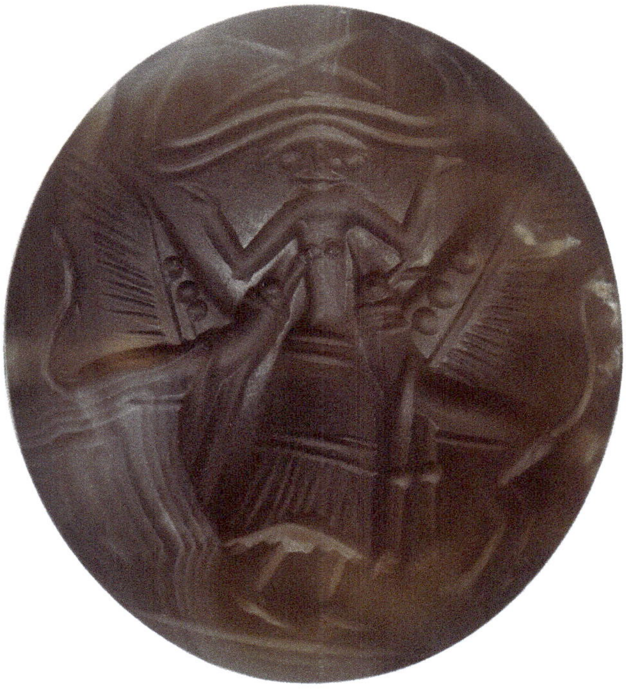

Fig. 12. Onyx gem from Knossos with a High Priestess-Goddess. Archaeological Museum of Herakleion. D.: LMII, c. 1450-1400 BC. (Photo: Archaeological Museum of Herakleion © HMC-ARF)

1450-1400 BC). (CMS II/3, no 63; Gimbutas 1989, 272, fig. 426; Haarmann 1991, 88, fig. 38 below left; Sakellarakis 1994, fig. p. 85 under right; Sapouna-Sakelaraki 1994, no. 82, fig. 82). It depicts a woman with the head and eyes of an insect, probably a bee or a butterfly; the identification is difficult. She is flanked by griffins (or by winged hounds according to other scholars); between the double bull horns ("snake frame") above her head is a double-axe-shaped butterfly or an astral symbol. Here, she is represented in her epiphany as Mistress of the Animals (on the Master and Mistress of Animals see: Christou 1968; LIMC II/1,2, 1984, 739 f.; Spatz 1962, 8-22; Tamvaki 1974, 287-91. Compare also Barclay 2001, 378-81, pl. CIIIa, CIVe; Crowley 1989, 34-39). It is a "performed or staged epiphany", namely a person (priestess or a member of the king-family) appears in the role of the Goddess (Hägg 1986, 56, nt. 89, 59; Liveri 2008, 5-11; Niemeier 1986, 74-77); the monster or the fabulous animals maybe were a part of the decoration. However, Tamvaki (1989, 263-70, fig. 5) underlines the "aniconic rendering of the human face with no indication of features" in the Aegean glyptic and she associated this with the limited space. Hooker (1992, 101) has a similar view—he believes, e.g., that the female heads on the ring of Isopata are represented as being in a hurry. Nevertheless, there are examples with treatment of personal features in the Minoan glyptic, e.g., the priests (CMS VIII, 110a-c). Morgan (1995, 136-37, fig. 4) emphasizes the frontal face of the Goddess on the onyx gem as being unique in Aegean art and she associates it with Death symbolism. However, even if the identification of the High Priestess-Goddess as insectomorphic is correct, this is unique and does not demonstrate the cult of a Bee-Goddess on Crete as Gimbutas and Haarmann argue. Later scholars explain the Minoan bee representations as the attribute and appearance of a water-divinity as well. Marconi, Picard, Mirié and others as well argue for the existence of a Bee Goddess in Minoan Crete. Graves (1975, 82.6) believes that Cer *(Κήρα)* was the Cretan Bee Goddess; a goddess connected with the wax as her name shows.

Jewellery in the shape of bees as burial offerings

However, the two bees in the lovely gold pendant from the funeral complex of Chrysolakkos by Mallia (Fig. 9), now in the Archaeological Museum of Herakleion (MM III, *c.* 1700 BC), are represented naturalistically. They are face-to-face and flank a honeycomb or pollen ball according to Kitchell. The granulated ball between their feet and another smoother smaller one between their mouths have been discussed by many scholars. Some of them, such as Warren, identify the insects as "wasps sucking a drop of honey", others such as Marinatos and Spanakis see "two hornets holding a granulated honeycomb". However, most scholars argue that honey bees are represented; in addition Kitchell distinguishes the balls as a honey drop (upper ball) and a pollen ball (lower ball). (Crane 1983, 217-18, fig. 259; Higgins 1957, 45,47, pl. 15e; Marinatos-Hirmer 1973, fig. 13; Sakellarakis 1994, p. 71, no 559, fig. p. 70; Sapouna-Sakellaraki 1994, no 65, fig. 65. On the theories of bees or wasps see: Davaras 1984, 82 f.; Hood 1976, 59-72; Kitchell 1981, 9-15, pl.1). The pendant is connected with the royal family of Mallia and must have a funeral character as symbol of life and death, renewal and rebirth (Dietrich 1974, 120; Hood 1976, 72; Richards-Mantzoulinou 1979, 77 f.). Higgins (1957, 57) associates another golden bee of the British Museum (British Museum Catalogue Jewellery, pl. xiv, no 1239), with the so-called Aegina Gold Treasure, originating in Crete with this bee-pendant. He believes that both are Minoan funeral offerings from the same tomb and show similarities with the subject and also the character of the granulation and of the filigree, and must be dated as the other pieces of the treasure to 1700-1550 BC.

In Mycenaean sites on the Peloponnese bees as burial offerings have also been found, probably under Minoan religious influence: We can refer to six gold, flying, bees which have been found in the coupled tomb of Peristeria by Pylos (Museum of Pylos, LH II, *c.* 1500 BC). They must have been part of clothes decoration (Hood 1976, 66, fig. 1, 9; Marinatos 1962, 97, pl. 99a; Idem 1963, B, 100, pl. 121b). Similar small round gold foils with carved bees among other insects, e.g. stylized butterflies or chrysalises (Fig. 13), have also been excavated in Mycenae (Shaft grave III-Royal Circle A), now in Archaeological Museum of Athens (LH I, 1600-1500 BC). However, most scholars support the identification with the last insects (Deppert-Lippitz 1985, 30, fig. 10; Dietrich 1974, 120; Iakovidis 1994, 338, fig. 71; Marinatos-Hirmer 1973, 177, fig. 224: he writes of bees; Richards-Mantzoulinou 1979, 78). This group of burial finds shows "probably a tendency in Minoan and Mycenaean period to bury the dead with a symbol related to the bee (among other symbols as well) that perhaps would assure the resurrection". In addition some scholars explain the Mycenaean coupled tombs "as hives of the souls" and they see a similarity between them and the shape of a bee-hive (Dietrich 1974, 120, nt. 313; Forbes 1957, 87; Richards-Mantzoulinou 1979, 78; Zafiropulo 1966, 40). This theory has been rejected by Davaras (1984, 86, nt. 9) with the argument that no beehive has been found with a similar shape. Nevertheless, I tend to disagree. In connection with similar ideas/beliefs it could be probable. Whether one sees the bee as symbol of resurrection and renewal, why must one refuse similar symbolism for its house, the beehive?

The Bee Goddess, honey and wax in the Linear B script

Richards-Mantzoulinou (1979, 81 f., nt. 24) agrees with Masouridi (1976, 69-74) and has accepted a Potnia Melissa (a Mistress Bee) on the tablet Tn 316 of Pylos with Linear B script which has been read as "*Εχις Πότνια...Μέλισσα Πότνια*" and she has related it with a snake- and bee-cult. The text in both sides of this tablet refers to rituals and offerings to deities of fertility, so both (snake and bee) belong to this category of divinities. However, other scholars such as Davaras and Koraka do not agree and argue for the older reading according to Ventris and Chadwick

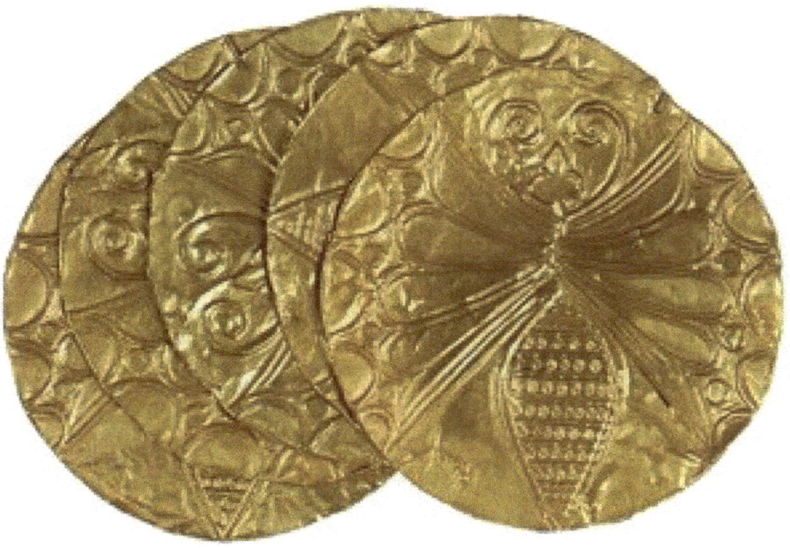

Fig. 13. Round gold foils with encarved butterflies or chrysalises from Mycenae. National Archaeological Museum of Athens. D.: LHI, 1600-1500 BC. Inv. no 1181(Photo: National Archaeological Museum of Athens © HMC – ARF)

(Davaras 1984, 80, nt. 11; Koraka 2001, 21, nt. 52; Ventris-Chadwick , 1959, 462 ff, 465 no 172 (Tn 316). Compare Boëlle 2001, 404, nt.1, 406, 408; Trümpy 2001, 414, 420).

Nevertheless, in general it seems that a connection between the snake- and bee-cult was probable, and this cult combines features of a chthonian and heavenly nature with a symbolism of the circle of life and death, although the archaeological evidence is insufficient to indicate it for the Minoan and Mycenaean times, e.g. the faience figurines of a snake goddess (MMIII, *c.* 1600 BC) and the ton model of a honeycomb with snakes wrapped around it in Archaeological Museum of Herakleion (Liveri 2008, 13 f., col. pl. III/1; Sakellarakis 1994, fig. p. 34, no. 8931, p. 39, fig. in p. 37, no. 63, 65; Sapouna-Sakellaraki 1994, 323, fig. 59, 60).

Information about the use of honey, e.g. as an offering to divinities, is contained in many Linear-B clay tablets from Knossos, Mycenae and Pylos. Worshippers offered honey, "honeyed" wine *(me-ri-ti-jo=μελίτιος οίνος)* or wax to gods such as Poseidon *(po-si-da-[o-ne]),* Eleuthia/ Eileithyia *(e-re-u-ti-ja)*, the Labyrinth Potnia= *(po-ti-ni-ja meri)* or to all gods *(pa-si-te-o-i/me-ri)*: (See the symbol for honey in the Linear B in Sakellarakis 1994, 49 and Ventris-Chadwick 1959, 223 ff. Regarding honey, honeycomb or wax as offerings compare also: Davaras (1984, 55-59, 84-85, 87-91, fig. 1); Melas (1999, 489 f.); and Moss (2005, 58-61). For later times see also Liveri (2000, 55, 61)). The word for honey maybe was me-ri (Gerard-Rousseau 1968, 83, nt. 17-21; also on meridumate and meriteu). Some Knossian Linear B Tablets have been found near reliefs showing bulls. As discussed above, however, this fact cannot be taken as evidence of a bull god or the worship of bull in the Minoan religion. The ideogram 168 of the Linear B is explained as a sign of the beehive, whereas 132 and 211 are signs representing honey amphorae or an "object" connected probably with honey or wax (Davaras 1984, 77). The significance of the honey in the Minoan Calendar is very interesting as well. According to Karl Kerényi (1961/62, 212 f.) the Minoans prepared a honey drink for their New Year feast/ceremony in a cult-cave connected with the rise of Sirius under Egyptian Calendar influence.

Conclusions

We conclude with the following remarks: The cult of the Bee Goddess as another form of the Great Goddess, Mother Earth or Goddess of Nature in Old Europe cannot be demonstrated with certainty. In the Neolithic, and later in the Bronze and Iron Ages (Minoan and Mycenaean cultures), the identification of most figures with a bee is uncertain and the arguments are not acceptable in all cases. Old theories of a Bee Goddess in the Minoan pantheon in connection with a potnia of plants and flowers or another Bee Potnia in the Mycenaean world, especially in Pylos according the reading of the Tablet *PY Tn 316* of Linear B, have been rejected.

We can imagine that the bee was a sacred insect as is shown by representations on seals, seal-impressions, lentoids, hieroglyphics, and funeral offerings as jewellery. It must have been connected with deities of Nature, vegetation, fertility, death, life and regeneration according to analogous religious beliefs, and probably after having been transformed slightly it was transferred into a later cultural period (Mycenaean) and survived as deities with different names, but with similar religious features. One can argue that a few Mycenaean archaeological finds show Minoan religious influences.

Acknowledgments: I express my thanks to museums for the dispatch of photographs and permission for their publication.

References

ATZENI, E. (1975/ 2004) Laconi – "Statua-Menhir" di Genna e Arrele. In G. Murru (ed.), *Enrico Atzeni, La scoperta delle Statue-Menhir: Trent'anni di ricerche* archeologiche nel territorio di Laconi, 20-33. Cagliari, CUEC.

ATZENI, E. (1981/2004). Le Statue-Menhir del Sarcidano. In G. Murru (ed.), see above, 69-73.

ATZENI, E. (1982/2004). Menhirs antropomorfi e Statue-Menhirs della Sardegna. In G. Murru (ed.), see above, 79-105.

ATZENI, E. (1989/2004). Megalitismo e Arte. In G. Murru (ed.), see above, 157-172.

BARCLAY, A. (2001). *Mastery of Animals: Reception of an Oriental Image in Archaic Greece.* PhD thesis, University of Toronto.

BOËLLE, C. (2001). Po-ti-ni-ja: Unité ou pluralité?, in R. Laffineur and R. Hägg (eds.), *POTNIA: Deities and Religion in the Aegean Bronze Age*, Proccedings of the 8th International Aegean Conference in Göteborg 12-15 April 2000. *Aegaeum* 22, 403-409. Liège, Université de Liège/Austin.

CAIN, C. D. (2001). "Dancing in the Dark: Reconstructing a narrative of Epiphany on the Isopata Ring". *American Journal of Archaeology* 105, 27-49.

CAMERON, D. O. (1981). *Symbols of Birth and Death in the Neolithic Era.* London, Kenyon-Deane.

CATALOGUE EXHIBITIONS (2003). *The Bull in the Mediterranean world. Myths and Cult.*, Barcelona-Athens, Museu d'Historia de la Ciutat de Barcelona and Benaki Museum.

CATALOGUE EXHIBITIONS (2006). *Η μορφή της αρχής. Καλλιτέχνες του εικοστού αιώνα και η Αρχαία Ανατολική Μεσόγειος.* Athens, Museum of Cycladic art.

CATALOGUE EXHIBITIONS (2007). *"Vor 12.000 Jahren in Anatolien. Die ältesten Monumente der Menschheit".* Karlsruhe, Theiss. Badisches Landesmuseum.

CATALOGUE EXHIBITIONS (2008). *Zeit der Helden. Die „dunklen Jahrhunderte" Griechenlands 1200-700 v. Chr.* Karlsruhe, Primus. Badisches Landesmuseum.

CAUVIN, J. Greek transl. by Preve S. (2004). Naissance des divinités - Naissance de l´agriculture (La Révolution des Symboles au Néolithique). Herakleion, Crete University Press.

CMS (1989). *Corpus der Minoischen und der Mykenischen Siegel.* Berlin, Gebrüder Mann

CRANE, E. (1979). *Honey (A comprehensive Survey.* London, Heinemann.

CRANE, E. (1983). *The Archaeology of Beekeeping.* London, International Bee Research Association.

CHRISTOU, Chr. (1968). *Potnia Theron.* Thessaloniki.

CROWLEY, J. L. (1989). *The Aegean and the East. An Investigation into the Transference of Artistic Motifs between the Aegean, Egypt, and the Near East in the Bronze Age.* Jonsered, Paul Åströms.

CUTTING, M. (2007). Wandmalereien und –reliefs im anatolischen Neolithikum. Die Bilder von Çatal Höyük. Catalogue Exhibitions, see above, 112-122.

DAVARAS, C. (1984). Μινωικό κηριοφόρο πλοιάριο της Συλλογής Μητσοτάκη. *Archaiologike Ephimerida*, 56-92.

DEPPERT-LIPPITZ, B. (1985). *Griechischer Goldschmuck.* Mainz, Von Zabern.

DICKINSON, O. (1994). *The Aegean Bronze Age.* Cambridge, University Press.

DIETRICH, B. C. (1974). *The Origins of Greek Religion.* Berlin, De Gruyter.

DOUMAS, C. (1983). *Thera: Pompei of the ancient Aegean.* London, Thames and Hudson.

EVANS, A. (1909). *Scripta Minoa: the written documents of Minoan Crete, with spezial reference to the archives of Knossos*, vol. 1. Oxford, Clarendon Press.

FORBES, R. J. (1957). *Studies in Ancient Technology*, vol. V, 78-97. Leiden, E. J. Brill.

FOSSATI, A., JAFFE, L. and SIMõES DE ABREU, M. (1991). *Durch die Zeiten eingemeisselt: Die Felsbilder des Valcamonica.* Valcamonica Preistorica, vol. 3. Milano, Litho Delta.

GARFINKEL, Y. (2003). *Dancing at the Dawn of Agriculture.* Austin, Texas University Press.

GERARD-ROUSSEAU, M. (1968). *Les mentions religieuses dans les tablettes mycéniennes.* Rome, Ateneo (Serie Incunabula Graeca n. XXIX)

GIMBUTAS, M. (1974). *The Gods and Goddesses of Old Europe.* London, Thames and Hudson.

GIMBUTAS, M. (1989). *The language of the Goddess.* San Francisco, Harper and Row.

GIMBUTAS, M. (1991). *The Civilization of the Goddess.* San Francisco, Harper and Collins.

GODART, L. (2001). La Terre Mère et le monde Égéen, in R. Laffineur and R. Hägg (eds.), see above, *Aegaeum* 22, 463-66.

GOLAN, A. (2003). *Prehistoric Religion (Mythology-Symbolism).* Jerusalem, Ariel Golan.

GRAVES, R. (1960). *The Greek Myths,* vol. 1, 2. London, Penguin.

HAARMANN, H. (1989). Writing from Old Europe to ancient Crete, *Journal of Indo-European Studies* 17, 251-75.

HAARMANN, H. (1991). *Universalgeschichte der Schrift.* Frankfurt/New York, Campus.

HAARMANN, H. (1996). *Early Civilization and Literacy in Europe (An Inquiry into Cultural Continuity in the Mediterranean world).* Berlin, De Gruyter Mouton.

HÄGG, R. (1986). Die göttliche Epiphanie im minoischen Ritual. *Athenische Mitteilungen* 101, 41-62.

HIGGINS, R. A. (1957). The Aegina treasure reconsidered. *British School of Athens* 52, 42-57.

HISTORY OF THE GREEK NATION (1971): (Encyclopaedia: Istoria Ellenikou Ethnous), Athens, Ekdotike Athenon S. A.

HOOD, S. (1976). The Malia Gold Pendant: Wasps or Bees?, In Fr. Emmison and R. Stephens (eds.), *Tribute to an antiquity: essays presented to Marc Fitch by some of his friends,* 59-72.London, Leopard's Head Press.

HODDER, I. (2007). *Çatal Höyük*. In Catalogue Exhibitions, see above, 124-25.

HOOKER, J. (1992). Early Balcan "Scripts" and the ancestry of Linear A. *Kadmos* XXXI, 2, 97-112.

HUTTON, R. 1997. The Neolithic great goddess: A study in modern tradition. *Antiquity* 71, 91-99.

JOHNSON, B. (1981). *Lady of the Beasts. Ancient Images of the Goddess and her Sacred Animals*. San Francisco, Harper and Row.

IAKOVIDIS, Sp. (1994). *Μυκηναϊκή τέχνη, in: ΕΛΛΗΝΙΚΗ ΤΕΧΝΗ. Η αυγή της ελληνικής τέχνης*. Athens, Ekdotike Athenon S. A.

KAKRIDIS, I. (ed.) (1986). *Ελληνική μυθολογία*, v. 2, 3. Athens, Ekdotike Athenon S. A.

KARAGEORGHIS, V. (1971). Notes on some cypriote priests wearing bull-masks. *Harvard Theological Review* 64, 261-263.

KERÉNYI, K. (1961-62). Licht, Wein, Honig: Die Frage nach dem minoischen Festkalender. In First International Cretan Congress. *Kretika Chronika* 15-16, 200-14.

KITCHELL, K. F. (1981). The Mallia "wasp" pendant reconsidered. *Antiquity* LV, 9-15.

KORAKA, K. (2001). A day in Potnia's life. Aspects of Potnia and reflected "Mistress" activities in the Aegean Bronze Age, in R. Laffineur and R. Hägg (eds.), see above, *Aegaeum* 22, 15-26.

LIMC: Lexikon Iconographicum Mythologiae Classicae VI/1 (1992) 444-446.

LIVERI, A. (2000). *Εικαστική απεικόνιση αρχαίων ελληνικών και λατινικών μύθων σχετικά με τη μέλισσα και τα προϊόντα της,* in: Acts «Η μέλισσα και τα προϊόντα της». 6th Workshop in Nikiti 12-15 September 1996. Athens, Cultural Technological Foundation ETVA.

LIVERI, A. (2008). Representations and Interpretations of Dance in the Aegean Bronze Age. *Athenische Mitteilungen* 123, 1-38.

MAHLSTEDT, I. (2004). *Die religiöse Welt der Jungsteinzeit.* Darmstadt, Wissenschaftliche Buchgesellschaft.

MARCONI, M. (1940). Melissa, dea cretese. *Athenaeum* 18, 164-178.

MARINATOS, N. (1993). *Minoan Religion (Ritual, Image and Symbol)*. South Carolina, University Press.

MARINATOS, N. (1994). *"The Export Significance of Minoan Bull Hunting and Bull Leaping Scenes"*. Ägypten und Levante IV, 89-93.

MARNATOS, N. (1999). Bull Hides as Dadoes and Emblems of prestige in Creto-Mycenean Palaces, in P. Betancourt, V. Karageorghis, R. Laffineur and W.-D. Niemeier (eds.), *MELETHMATA:Studies in Aegean Archaeologyp presented to Malcolm H. Wiener as he enters 65th Year. Aegaeum*, 20, 481-83. Liège, Université de Liège/Austin.

MARINATOS, Sp. 1962. Ανασκαφαί Πύλου. *Praktika Archaeologikes Etaireias*, 90-98.

MARINATOS, Sp. 1963. Ανασκαφαί Πύλου. *Arhaeologikon Deltion* 18, 99-100.

MARINATOS, Sp. and HIRMER, M. (1973). *Kreta, Thera und das Mykenische Hellas*. München, Hirmer.

MASOURIDI, N. A. (1976). Η μινωική γραφή ε„ς τάς πινακίδας της Πύλου, Athens.

MATZ, Fr. (1961-62). Minoischer Stiergott. in Acts of the First International Cretan Congress, *Kretika Chronika* 15-16, 215-223.

MELAS, M. (1999). The Ethnography of Minoan and Mycenaean Beekeeping, in P Betancourt, V. Karageorghis, R. Laffineur and W.-D. Niemeier (eds.), *see above, Aegeum* 20/II, 485-490.

MELINI, R. (2008). "Doing" musik without "making" musical instruments: A prehistoric, actually contemporaneous, concept, *in the session "Prehistoric technology: cognition and the act of artefact production*", Abstracts of the 14th Annual Meeting of the European Archaeologists 16-21 September 2008 in Malta, 168 (Awaiting publication).

MELLAART, J. (1967). *Çatal Hüyük. Stadt aus der Steinzeit*. Bergisch Gladbach, Gustav Lübke.

MELLAART, J. (1970). *Excavations at Hacilar.* Edinburgh, University Press.

MESKEL, L. (1995). Goddesses, Gimbutas and "New Ag" Archaeology. *Antiquity* 69, 74-86.

MILOJČIĆ-VON ZUMBUSCH, J. and MILOJČIĆ, V. (1971). *Die deutschen Ausgrabungen aus der Otzaki-Magula in Thessalien,* part I, II. Bonn, Habelt.

MIRIÉ, S. (1979). *Das Thronraumareal des Palastes von Knossos*. Bonn, Habelt.

MORGAN, L. (1995). Frontal face and the Symbolism of Death in Aegean Glyptik. In I. Pini and J.-Cl. Poursat (eds.), "*Sceaux minoens et mycéniens*". IV Symposium International 10-12 Sept. 1992 in Cremont-Ferrand. *Corpus der Minoischen und Mykenischen Siegel*, Beiheft. 5, 135-149. Berlin, Gebrüder Mann.

MOSS, M. L. (2005). *The Minoan Pantheon*. Oxford, BAR International Series 1343.

NIEMEIER, W.-D. (1986). Zur Deutung des Thronraumes im Palast von Knossos. *Athenische Mitteilungen*, 101, 64-95.

NIEMEIER, W.-D. (1989). Zur Ikonographie von Gottheiten und Adoranten in den Kultszenen auf minoischen und mykenischen Siegeln. In I. Pini (ed.), "*Fragen und Probleme der bronzezeitlichen ägäischen Glyptik*". 3rd International Seal-Symposium in Marburg 5-7 Sept. 1985. *Corpus der Minoischen und Mykenischen Siegel*, Beiheft 3, 163-186. Berlin, Gebrüder Mann.

NILSSON, M. P. (1927/r.1950). *The Minoan-Mycenaean Religion and its Survival in Greek Religion.* Lund, Gleerup.

OTTO, Br. (1985). *Die verzierte Keramik der Sesklo- und Diminikultur Thessaliens*. Mainz am Rhein, Von Zabern.

PACE, A. (2006). *The Tarxien Temples*. Malta, Heritage.

PAPAEFTHYMIOU-PAPANTHIMOU, A. (1997). *Τελετουργικός καλλωπισμός στο προϊστορικό Αιγαίο*. Thessaloniki, Vanias.

PAPAPOSTOLOU, I. A. (1977). Τα σφραγίσματα των Χανίων: Συμβολή στη μελέτη της μινωικής σφραγιδογλυφίας. Athens, Library of the Athens Archaeological Society.

PICARD, Ch. (1948/repr. 1972). *Les religions préhelléniques*. Paris, Université de France Press.

RENFREW, C. (1972). *The Emergence of Civilization. The Cyclades and the Aegean in the Third Millennium B.C.* Frome and London, Methuen and CO LTD.

RICHARDS-MANTZOULINOU, E. (1979). «Μέλισσα Πότνια». *Athens Annuals of Archaeology* XII/1, 72-92.

RUTKOWSKI, B. (1981). *Frühgriechische Kultdarstellungen*, Athenische Mitteilungen, Beiheft 8. Berlin, Gebrüder Mann.

SAKELLARAKIS, I. A. (1994). *Μουσείο Ηρακλείου (Εικονογραφημένος οδηγός)*. Athens, Ekdotike Athenon S. A.

SAPOUNA-SAKELLARAKI, E. (1994). *Μινωική τέχνη, in: ΕΛΛΗΝΙΚΗ ΤΕΧΝΗ. Η αυγή της ελληνικής τέχνης*. Athens, Ekdotike Athenon S. A.

SHARON, S. (2006). *The National Museum of Archaeology. Valetta. The Neolithic Period.* Malta, Heritage.

SCHLAGER, N. (1989). Minotauros in der ägäischen Glyptik?, in: I. Pini (ed.), see above, *Corpus der Minoischen und Mykenischen Siegel,* Beiheft 3, 224-237.

SPATZ, E. (1962). *Das Wappenschild des Herren und der Herrin der Tiere in der minoisch-mykenischen Kunst*. PhD thesis, University of Munich.

TAMVAKI, A. (1989). The Human Figure in the Aegean Glyptic of the Late Bronze Age: Some Remarks, in I. Pini (ed.), see above, *Corpus der Minoischen und Mykenischen Siegel*, Beiheft 3, 259-273.

TAMVAKI, A. (1974). The seals and sealings from the citadel house area: a study in Mycenean Glyptic and Iconography. *British School of Athens*, 69, 259-93.

THOMAS, C. G. and WEDDE, M. (2001). Desperately seeking Potnia. In R. Laffineur and R. Hägg (eds.), see above, *Aegaeum* 22, 3-13.

TRÜMPY, C. (2001). Potnia dans les tablettes mycéniennes: Quelques problèmes d'interpétation. In R. Laffineur and R. Hägg (eds.), see above, *Aegaeum* 22, 411-421.

TRUMP, D. H. (2005). *Malta. Prähistorische Zeit und Tempel*, transl. S. Tröster. Florenz, Midsea.

VENTRIS, M. and CHANDWICK, I. (1959). *Documents in Mycenean Greek.* Cambridge, Cambridge University Press.

YULE, P. (1981). *Early Cretan Seals: A Study of Chronology.* Mainz am Rhein, Von Zabern.

ZAFIROPULO, J. (1966). *Mead and wine: A History of the Bronze Age in Greece.*London, Widgwick and Jackson.

A Case of Prehistoric Incubatio: the Cult Site of Civitaluparella, Chieti, Italy

Tomaso Di Fraia

Dipartimento di Scienze Archeologiche, Università di Pisa
Via S. Maria 53, 56126 Pisa, Italia
Correspondence: Via Francesco Redi 11, 56124 Pisa
e-mail: tom.difraia@libero.it

***Abstract**: The recent find of the Pastine site, near the Civitaluparella township, seems to be a very important prehistoric cult place. The main features are: (1) a rock face painted with an anthropomorphic female figure and three geometrical signs; (2) at the base of this wall, a long flattish stone slightly sloping with three engraved signs (two single crosses and a cross inside a circle) on its top surface; (3) two other crosses are engraved on the rock wall, close to the stone, in a small strongly-smoothed area; (4) the part of the stone close to the wall is also smoothed for its entire length. All these elements show that the large stone was used as a bed, in which a person lying down could touch with the right hand just the portion of wall with the two engraved crosses while simultaneously viewing the painted figures.*

The context fully corresponds to the conditions already known for the ritual called incubatio by the Romans. This ritual required the permanence of a person on a rock bed to obtain some benefit (like a recovery from illness) from some supernatural power.

***Key words:** Civitaluparella (Italy), rock painting and engraving, symbolic signs, rock bed, incubatio, lithotherapy.*

***Riassunto**: Il sito delle Pastine, recentemente scoperto presso la cittadina di Civitaluparella, si è rivelato un importante luogo di culto preistorico. Le principali caratteristiche sono: 1) una parete rocciosa con una figura femminile e tre forme geometriche dipinte; 2) alla base della parete, un grande masso liscio e lievemente inclinato con tre segni incisi (due croci singole e una croce inscritta in un cerchio); 3) altre due croci sono incise sulla parete rocciosa, vicino al masso, in una piccola area fortemente consumata; 4) anche la parte del masso verso la parete è consumata per tutta la sua lunghezza. Tutti questi elementi dimostrano che il masso era usato come un letto, da cui una persona sdraiata poteva toccare con la mano destra proprio la porzione di parete con le due croci incise e contemporaneamente vedere le figure dipinte.*

Il contesto corrisponde perfettamente alle condizioni che noi conosciamo per il rito chiamato incubatio dai Romani. Il rito richiedeva la permanenza di una persona sopra un letto di pietra al fine di ottenere qualche beneficio(soprattutto guarigioni da malattie) da parte di qualche entità soprannaturale.

***Parole chiave:** Civitaluparella (Italia), pitture e incisioni su rocce, figure simboliche, letto di pietra, incubatio, litoterapia.*

The discoveries

In February 2006, in the Municipality of Civitaluparella, locality Pàstine, Prof. Aurelio Manzi (Di Fraia and Manzi, in press) discovered some paintings and engravings (afterwards called the "Parete Manzi") on a rock wall ranging in direction east to west and jutting out in some places. The site is located on the left of the Sangro river, at approximately 550m above sea level. In a sub-vertical sequence four figures of red colour appeared (Fig. 1 and 2).

The first from the bottom represents a shield within which a cross is inscribed. The second figure, the largest (nearly 300mm high), is anthropomorphic, with a leg represented by a vertical axis that rises to the height of the flanks forming a sub-circular shape, while the head is rhomboid and the arms are constituted by two cruciform modules. The third figure, shifted a little to the right, is a rectangle cut by an irregular diagonal. The last figure differs clearly from the others, because it has been obtained tracing a thin line the result of rubbing a solid dye and not by an application with a paint-brush: it is an irregular quadrilateral cut by an irregular line, while two "V" shaped segments protrude at an angle. Near this angle some sub-vertical hyphens (two or three millimetres long) are engraved so as to render a crushed ellipse. At a distance of 180mm, on the left, another group of vertical hyphens is cut by two narrow vertical engraved lines 40 to 50mm long. The perimeter of the quadrilateral has been redrawn in irregular fashion with a thin engraved line.

On the right of these paintings, towards the bottom, two crosses have been created by tapping, probably by indirect percussion with a pointed chisel; the portion of rock under the inferior cross is very worn away, both in a elliptic flat area and on some corners (Fig. 3). An inspection carried out by Prof. Manzi and myself in August 2007 allowed us to recognise, alongside the base of the wall, just under

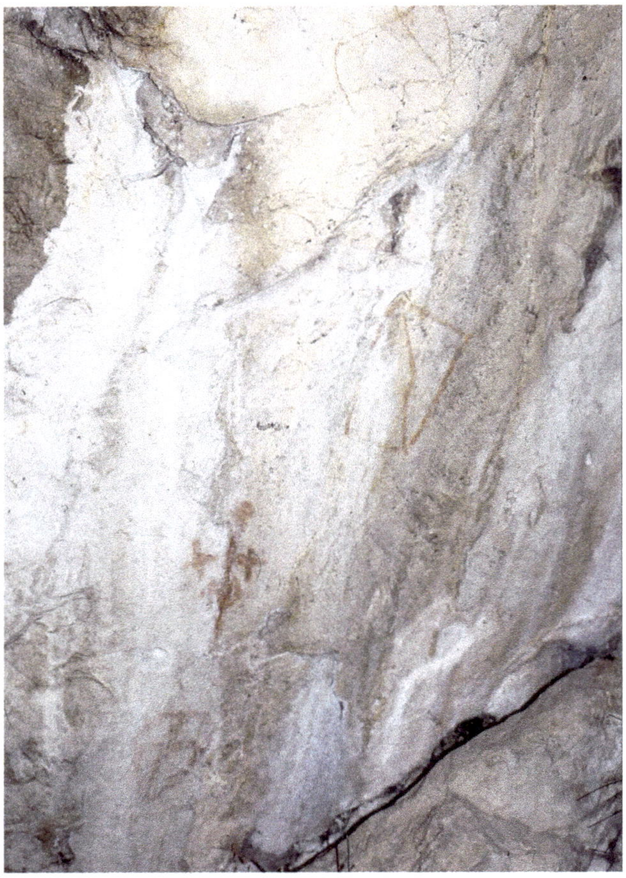

Fig. 1. Civitaluparella, Parete Manzi: Sequence of red figures (Photograph T. Di Fraia).

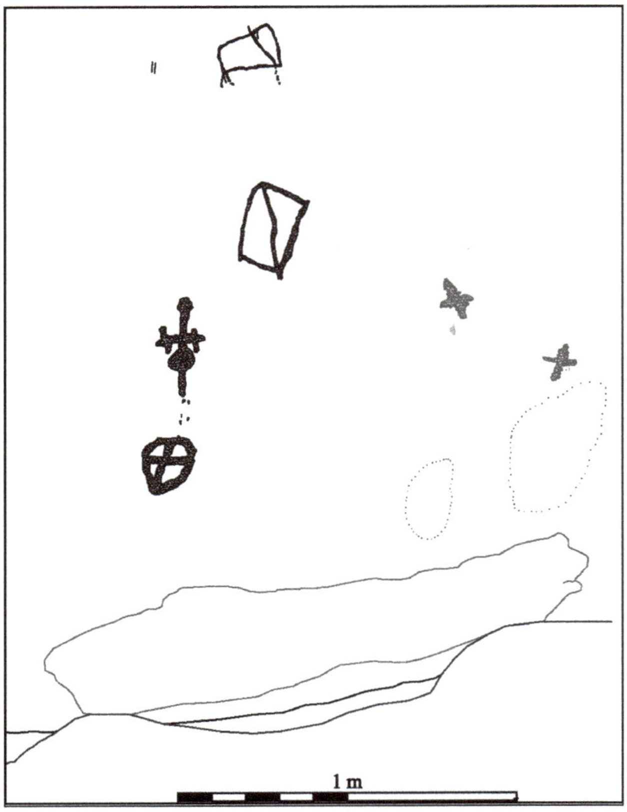

Fig. 2. Civitaluparella, Parete Manzi: Red figures and two engraved crosses; on the right the dots delimit two smoothed areas (Plotting by M. Colombo and M. Serradimigni).

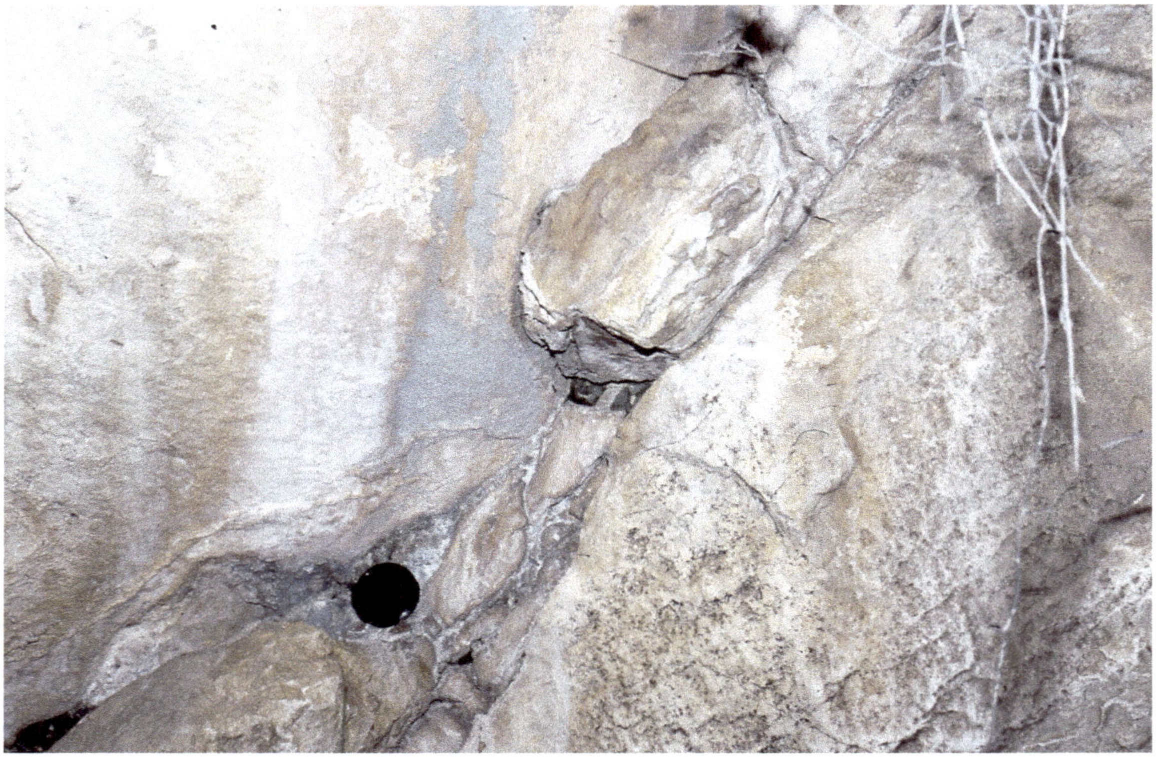

Fig. 3. Civitaluparella, Parete Manzi: Two engraved crosses (showing up as white, left of centre top and right of centre) and the smoothed wall (Photograph T. Di Fraia).

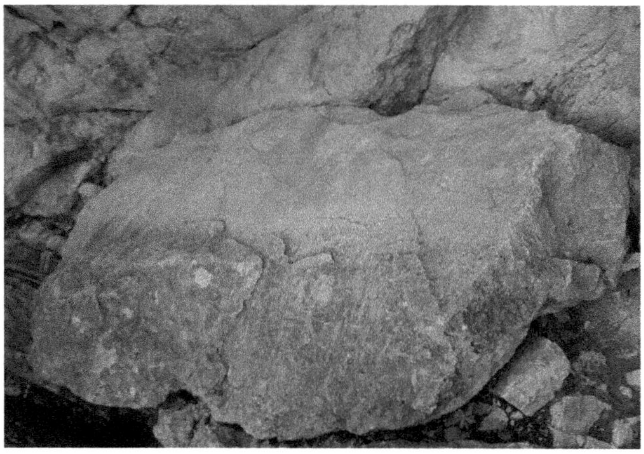

Fig. 4. Civitaluparella, Parete Manzi: Angled view of the large stone ("bed"), strongly worn in the inside part and on the right sharp (Photograph T. Di Fraia).

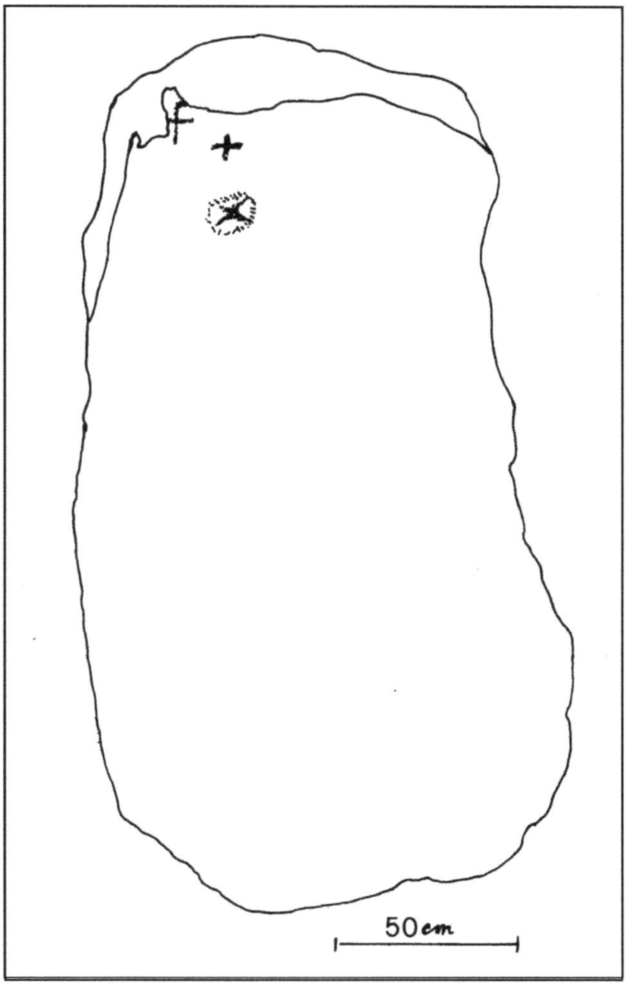

Fig. 5. Civitaluparella, Parete Manzi: Upper surface of the stone "bed" with engravings (Drawing by G. Giannessi).

the vertical painting sequence, a large stone (300-350mm thick, 2.34m long, 1.38m wide) lightly inclined from E to W. Its upper surface is saddled and appears to be worn away for the whole length and for about two-thirds of the width towards the rock wall. The wearing both on the wall (under the inferior engraved cross) and on the large stone, eliminating or reducing the roughness caused by numerous little bumps, has produced a regularisation and smoothing of the surface, that still conserve today a clearer coloration because of not being attacked by micro-organisms that do not take root on the nearly shiny parts (Fig. 4). Moreover the sharp of the eastern extremity of the large stone is particularly worn (Fig. 4, on the right).

Near such extremity on the upper surface of the stone three figures are engraved: an irregular circle with two not perpendicular diameters and two simple crosses, one of which has been slightly mutilated by a fracture of the cliff (Fig. 5). The circle is obtained with a series of little cuts (three to eight mm long) disposed in irregularly radial way, while the crosses, besides similar cuts, show also signs of point percussions.

A short archaeological campaign in August 2008 led us to discover other figures: (1) in the western part of the wall two crosses painted in red and some red spots (Fig. 6); (2) a bit to east a dagger-shaped figure, also painted red; (3) an engraved cross; (4) near the eastern edge of the wall a series of crosses and a probable anthropomorphic image created by digging a track or by percussion (Fig. 7).

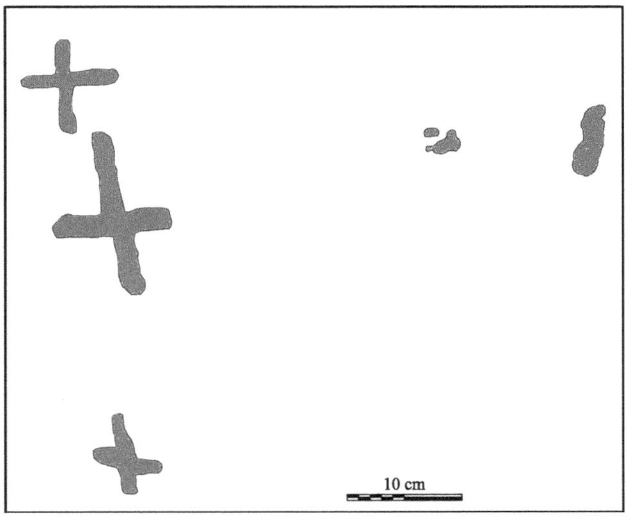

Fig. 6. Civitaluparella, Parete Manzi: Red painted crosses (Drawing by M. Colombo and M. Serradimigni).

The area near the base of the wall in correspondence of the figures and all around and under the large stone have been excavated, but we have not found any object of archaeological interest, nor any trace of anthropic or cultural deposit except a few fragments of wood and reeds partially burned, small charcoal fragments, animal excrements, a fragment of cloth and a modern bone button.

It is probable that the site has been visited in relatively

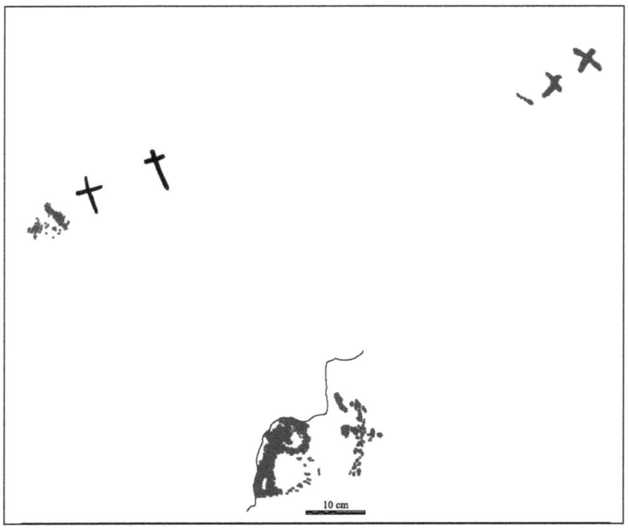

Fig. 7. Civitaluparella, Parete Manzi: Engraved crosses and an anthropomorphic form (Drawing by M. Colombo and M. Serradimigni).

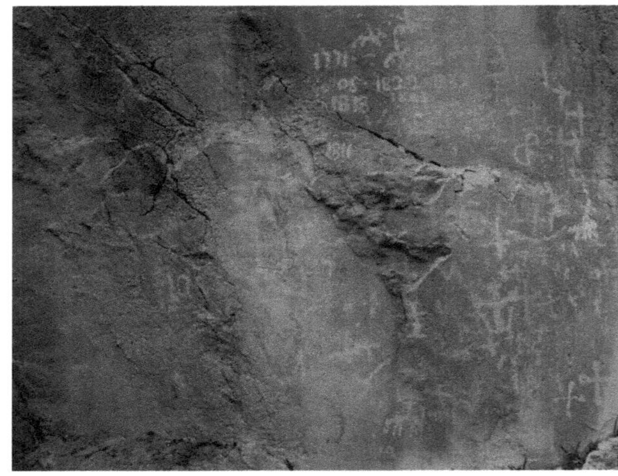

Fig. 8. Civitaluparella, Parete Di Cicco: Central panel with various engraved figures (Photo M. Serradimigni).

recent times by shepherds who have lit fires, as some black tongues on the base of the wall seem to testify. The absence of prehistoric finds is particularly strange if one considers that the area of interest is constituted by an irregular rock base, with many tracks, depressions and unevenness that would have certainly retained any possible materials of anthropic origin. Such absence gives rise to at least two problems. First of all it does not allow to date with precision the frequentation of the site; secondly, because an intense frequentation, however, is testified by the aforesaid traces of wear, how can one explain the absence of archaeological finds?

One could propose a hypothesis that the site was part of a more complex system that included some points in which a physical permanence only was previewed and others in which rituals could be carried out with the use of tools and with a possible accumulation of anthropic remains. This hypothesis seems reasonable in the light of a successive discovery carried out in the same territory. In 2008 Mr. Antonino Di Cicco, resident in Civitaluparella, announced to Aurelio Manzi the presence of a rich series of rock engravings (many figures, dates and a piece of writing) concentrated in a short space (approximately two metres wide) of a limestone wall (called "Parete Di Cicco") partially jutting out, over 100m long and approximately 20m high, in the southern slope below Civitaluparella, and approximately 750m above sea level, not much distant from the Prezzacchie spring. Figures and dates are obtained mostly with the same technique and display similar surfaces and patinas; some figures represent men on horseback and with dogs, there are many crosses, while the dates have as extremes 1771 and 1839 (Fig. 8). However, besides such figures that are likely nearly all of modern age, we have discovered in other points of the wall some red ochre paintings, other paintings in black, and many scattered engravings that seem prehistoric. Proceeding upstream (i.e. toward NE) from the down part (SW) of the wall, we find in succession these elements, distributed along 70m approximately: a small tapped cross; a complex grid figure painted black; two rectangles with diagonals, one above the other, and beneath these a crossed circle, all painted in red (Fig. 9); a series of black signs mainly tied that evoke cursive writing; a great panel of figures of men, animals, crosses and dates, on whose margins some tapped geometric figures seem to be more ancient; an engraved crossed circle; a series of black figures between which are two irregularly rhomboid ones; an engraved quadrangular figure; several black signs; two great complex figures painted red and difficult to read; and a tapped Latin cross. In this paper we limit ourselves to emphasize in particular the presence of the crossed circle, that we will discuss later on, and the two rhomboid figures that could allude to the vulva.

The Parete Manzi context

The only reasonable explanation for the traces of wear on the rock in the Parete Manzi context is a repeated series of two types of human activities. The first is that of lying on the "bed" of stone, so as to occupy the part (approximately two thirds of the width) near the wall, for all the length of the bed (Fig. 10); only such movements can explain the partial smoothing of the upper surface of the stone (the external part is not smoothed). The second type of movement has produced a strong smoothing of a much more limited area (about 300 mm wide) on the rock wall, a little above the "bed" and immediately under one of two engraved crosses, that one lower. It is likely that such rubbing could be produced with the right hand by a lying person, but it cannot be excluded in absolute that the same person or others could act by being in a different position and also with another part of the body.

In this initial phase of research we cannot be sure of the total

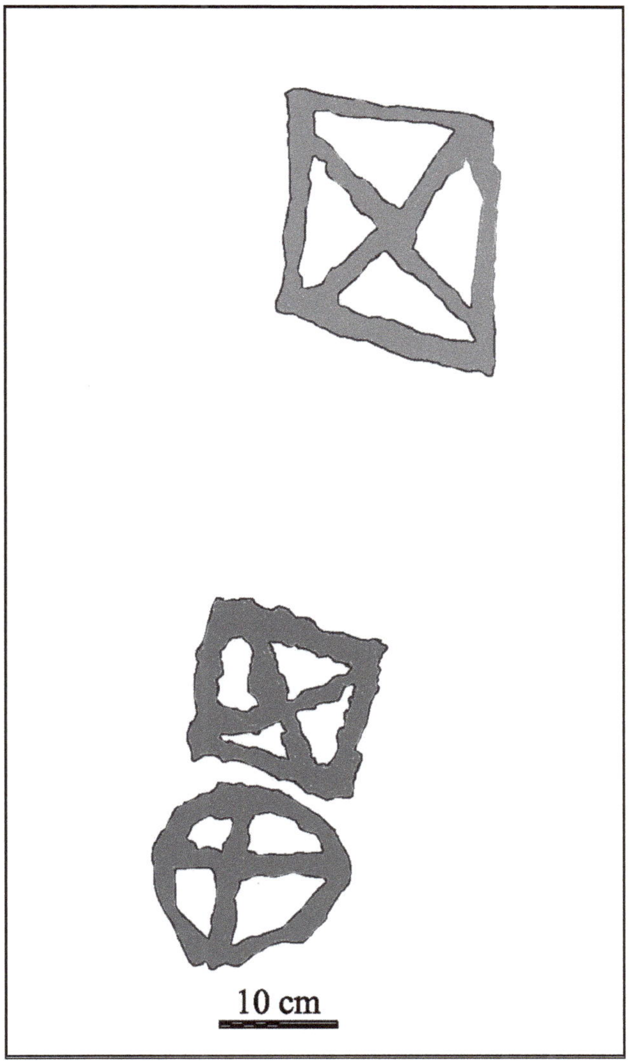

Fig. 9. Civitaluparella, Parete Di Cicco: Red painted figures (Drawing by M. Colombo and M. Serradimigni).

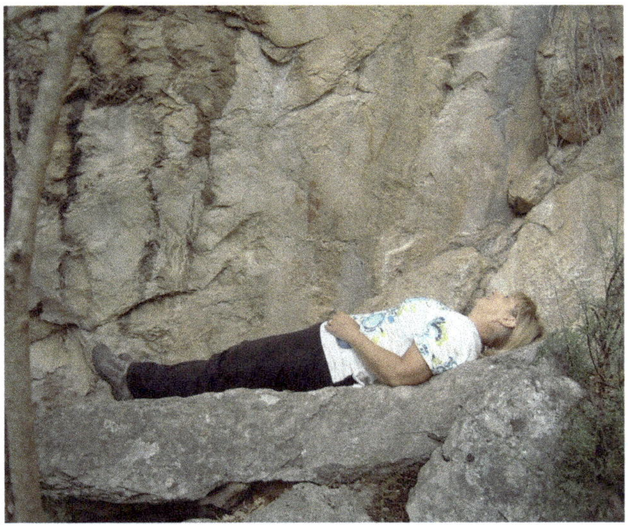

Fig. 10. Civitaluparella, Parete Manzi: Example of lying on the "bed" (Photo T. Di Fraia).

contemporaneousness between the presence of engraved crosses and the painted figures on the one hand and the rubbing of the wall on the other hand. Therefore we cannot exclude that the two engraved crosses have been added at a time posterior to the smoothing of the wall. Perhaps only a careful study of patinas and traces of wear will be able to give more precise indication.

Also, other parts of the low band of the wall, and especially of the rock that runs along the base, show signs of wear (particularly the smoothing of the sharp places). Such signs must be interpreted as the effect of repeated passages and consequent friction—perhaps of the feet, but others parts of the body cannot be excluded—against the rock.

Analogous traces of smoothing have been found on two stones near the eastern edge of the bed.

As for the symbols or figures overhanging the bed, which are perfectly visible by a person lying on it (Fig. 10), the most interesting is certainly the anthropomorphic one, drawn by following a substantially cruciform outline, with the vertical axis that widens in the median part with arms formed by two small cross modules. It seems, therefore, that the cross is not only isolated, but it represents the principle and the unifying element both of this figure and of the crossed circles. The anthropomorphic figure, although it is stylized, seems to refer to the feminine gender.

In my opinion, the only known practice that can explain the evidence formed by the position of the large stone and the immediate surrounding area is that of *incubatio*, which is the ritual lying down of persons on the stone, quite likely with a therapeutic aim, as our knowledge of this phenomenon in historical times suggests. Such a hypothesis, in the current state of the research, cannot exclude others that, however, can lead to the basic condition of lying down and to whatever can be derived from such a condition.

In this prospect, we could suppose also a frequentation of the site for initiatory purposes that perhaps previewed both a permanence on the stone and a particular track along the wall, as could be suggested by the traces of wear and smoothing on the rock, for many metres, to the base of the wall.

Finally we could think of a form of shamanic trance, as for example is assumed for the statuettes of women lying and/or sleeping and as found in some prehistoric sanctuaries in Malta. In this case the *incubatio* would be practised exclusively or mostly by a particular category of persons (shaman, medium, healer) and not by several members (in the condition of being patients) of a community, but either way it would be a ritual practice of considerable symbolic and social value.

The figure of the shaman could be present in every one of the assumed scenarios. The feminine kind of the painted anthropomorph could be correlated to the presence of a woman figure who physically (healer, medium, shaman) presided at the ritual, especially in the case of a therapeutic *incubatio*, or to a superior entity (Mother Goddess? or Earth Mother?) invoked for the fulfilment of the ritual. Regarding this point we can ask, what can have pushed a human group to choosing the site as a ritual place. In the current state of the our knowledge, because in the vicinity there is no spring, and the dripping of water on to the rock wall seems very modest, a direct relationship with water does not seems admissible.

What could have attracted the human interest seems therefore the presence both of a rock wall somewhat smooth and elevated, exposed to the south towards the Sangro river, and of an almost contiguous large stone having a particular shape. Such a hypothesis could correspond to a cultural attitude and a mentality that tried to read any kind of divine indication in the natural features of a place, so such features must be "understood" (that is correctly interpreted), respected and eventually marked and enriched to use them in the best way without violating the natural order. It can be marked that the visibility of the main group of red figures by the person lying on the "bed" is favoured from the jutting-out inclination of the rock wall at this point. Such localization of paintings is even more meaningful if it is considered that for the prehistoric rock paintings discovered in the province of Chieti (but also elsewhere) sub-vertical walls were generally preferred.

Someone has advanced the hypothesis that in some cases the paintings were realized generally at chosen points subject to dripping or a slight sliding of calcareous water and therefore to the formation of thin layers destined to veil or to obliterate the paintings. In the Parete Manzi this could be true only in partial measure, limited to the figures localized in the western part of the wall.

We have found strong traces of wear on the rock surface, also at other rock sites with paintings and engravings, in particular those of Fara–Colle Marino and of Lama-Fossato (Di Fraia Manzi, in press). At these two sites the rock base and a part of the wall (up to a height of nearly two metres in the niche at Colle Marino) that corresponds to the paintings are strongly smoothed. The site of Colle Marino shows another interesting feature: at two sides of the niche there are two large holes, apparently intentional, that could have served for positioning a beam or a horizontal table (Manzi and Manzi 2007, 26). Such support could be used as a bed for *incubatio*, as well as probable support for the persons who realized the paintings, not being reachable from the base of the wall. The wear of the rock could be explained as resulting from repeated contacts and rubbings carried out by frequent visitors to the site for ritual purposes.

Incubatio and lithotherapy

As is well known, the *incubatio* (in Latin it means literally: the condition of lying down) was a ritual practice diffused in the Greek and Roman world and connected particularly to the god Asklepios (*Aesculapius* for the Romans), whose most famous sanctuary was at Epidauros. The ritual previewed that the involved person slept in a sacred place, expressly predisposed and often equipped with stone beds, in order to obtain a vision, a dream, a revelation, or directly a recovery from some disease or inability, in particular feminine sterility. The possible effects of the *incubatio* were produced from autosuggestion processes, favoured by particular diets or fasts, by purifying baths, by the very discomforts of the stone on which a person was lying, or by the general environmental conditions (darkness, cold, particular sounds or noises etc), but probably the sleep was often induced by means of psychotropic substances. Some medicines had to be used sometimes or successively prescribed with a therapeutic aim, but they are cited very rarely in the classic sources.

Even in the modern and contemporary age—until the first half of the twentieth century—several forms of incubatio, lithotherapy and fertility cults are documented. In some cases, although under christianised forms, we can still recognize traces of conceptions and rituals that have their roots in ancient pagan cults. Sometimes explicit pagan references to the sexual and reproductive sphere have been in some measure disguised or counterfeited (Carabelli 1996). Also the Christian cults in recently-instituted sacred places often resort to practices that revive some traditional constants, in particular the rubbing of natural rocks and the use of spring water, as is done for example at Lourdes.

In Abruzzi the most ancient documentation of the *incubatio* has been discovered in the Grotta Di Ciccio, in Fucino. In this cave, excavated by Antonio M. Radmilli, six blocks of stone intentionally squared were brought to light and interpreted as beds for *incubatio* rituals (Radmilli 1956).

In Radmilli's excavation, and in several others previously and subsequently, many votive offerings were recovered, comprising phalli, uteri, breasts, children wrapped in bands, feet, hands, quadrangular tablets with human face in low relief—and various ceramic objects (vessels, cups, amphorae, balsamari, oil-lamps) that, together with some coins, allow us to date the use of the cave as a cult place from the seventh to the first century BC. In the Middle Ages a part of the cave was used as a Christian church dedicated to S. Felice (the current name of the cave is the result of fusing the name of this saint and the surname of an administrator of the city of Avezzano in the fourteenth century, Giovanni di Ciccio). Such christianisation of the cave is meaningful, as S. Felice, Pope from AD 269 to 274, celebrated on May 30. This is a moment traditionally dedicated to the rituals of fertility, while the Latin name (*Felix*) means "fecund", "prosperous". Another indication of cults related to prosperity is the "milk" ritual attested "… in the near church of S. Padre, obtained in the Discenderia Maggiore of the roman emissary of the Fucino—a sacred cavern where the marsican women collected objects previously dipped in a spring-water pool in order to ensure abundant flow of milk (Phoebonius 1668, II, 92)" (Grossi 2000).

Limited to the Abruzzi region, the practice of lithotherapy—that is, the tradition of rubbing on particular rocks, stones or walls the aching or sick parts of the body, with the aim of healing—still persists in various areas of the Abruzzi hills. These places are generally in caves used for the cult of Christianity or as centres for traditional hermitages. One site still frequented for this ancient practice is the hermitage of Sant'Onofrio on the Maiella in the Municipality of Serramonacesca (Province of Pescara). On the day of the festivity the faithful lie down and rub themselves in the "crib of Sant'Onofrio", which is a socket in the rock by now very worn, in the hope of recovering from abdominal pains and from fevers (Micati 1990).

Also on the Morrone mountain, in a wall that overrides the ruins of a temple dedicated to *Hercules Curinus*, there is a hermitage dedicated to Sant'Onofrio, beneath which is a cave dedicated to Pietro from Morrone. Here the faithful practice ritual rubbing of the aching parts of the body against the rocky wall. It also seems that in times past people practiced *incubatio*, sleeping in contact with the rock believed to be miraculous (Pansa 1924; Micati 1990). It is important to notice that not far from the hermitage there are two shelters with prehistoric paintings (de Pompeis 1993). In the hermitage of the saint Venanzio in the Municipality of Raiano (L'Aquila), pilgrims rub the sick parts of the body against the walls of the cave and in particular a socket believed to be the place where the saint used to rest (Fucinese 1995). Also in the gorges of S. Venanzio there is a group of prehistoric paintings (Mostacci 1997).

In the S. Michele cave (Municipality of Liscia, CH), in the saint's festivity day, the faithful rub themselves against the cave walls and drink ritually the water that gushes from inside the cavity (Manzi and Manzi 2004). Analogous forms of lithotherapy persist in other Abruzzi localities, among them the Gran Sasso massif, in the hermitages dedicated to S. Franco at Assergi (L'Aquila) and to S. Colomba at Isola del Gran Sasso (TE) territory (Micati 2002). It is important to remark that in some cases these ritual practices are not tied to devotion for particular saints, but to certain rock places, rendered even more famous by the presence of hermits or connection to certain saints, only because these had dwelled there for some time.

Also this element would seem to confirm the idea that it has not been the presence of particular saints to attribute sacredness and miraculous power to certain places, but it has been just these places to attract the popular cult and then the presence of saints—as an historical memory of long duration that could sink its roots in prehistory—had perpetuated their sacred character. The context of the Parete Manzi and the portions of strongly-smoothed rock at the sites of Fara-Colle Marino and Fara-Fossato supply a valid support for this hypothesis.

In fact these sites, although they are extraneous to the popular cult of the Christian tradition, probably because they were abandoned already in prehistoric times, have the two fundamental features that we find in the historical forms of lithotherapy and *incubatio*: namely, the presence of sanctifying elements (engravings and paintings) and evidence of lying and/or of repeated rubbings against the rock. Because the water has often played a key role in the selection and institutionalization of sacred places, the importance of the site of Fara-Colle Marino may also be related to the proximity of the massive springs of the Verde that flow just in front.

Hypothesis of chronology

Concerning the antiquity of the paintings and engravings of Parete Manzi and of a part of those of Parete Di Cicco, it seems that we can rule out a chronology of historical age for the following reasons:

1. In the area in question we do not know of any modern or historic rock painting in red;
2. As regards the typology of the rock figures in the Civitaluparella territory, the crossed circles and quadrilaterals with one or two diagonals are not documented by any historical representation painted or engraved. This is evident especially in the central panel of the Parete Di Cicco, where we have many variations of the Latin cross, but no crossed circles;
3. In the case of the Parete Di Cicco, the engraved figures are certainly modern (men and animals, and dates) and carried out mainly by a different technique (scraping or chipping almost tangentially to the surface) compared to the figures attributable to prehistoric times (chipping or chiseling generally perpendicular to the surface).

Once we have excluded a date of historical age, can we suggest a more precise chronological attribution and an interpretative proposal? As regards the technical aspects it is known that the red paintings are documented mainly in the Neolithic and in the Copper and Bronze Ages, while the black ones especially realte to the Iron Age, but it is only a general rule.

A more meaningful indication seems to come from the circle with an "x" engraved on the "bed" at the Parete Manzi. Indeed, this circle was obtained by juxtaposing a series of segments that appear to be carved with a chisel edge which must necessarily be made of metal, so this figure should be ascribed to the age of metals.

Unfortunately, the subjects of the rock paintings at the two sites presented here do not give us sufficient information to make comparisons with other rock-art contexts already dated, nor do they contain auto-dating elements like representations of tools, weapons or other special items that can be dated archaeologically.

Probably the most interesting subject is the crossed circle, i.e. a circle with four rays at right angles, which is present at both sites of the Civitaluparella territory. The simple cross, although widespread, is a figure little characterized to constitute a guiding element, even if it could be an extreme simplification of the same subject. If we now look to the overall archaeological record, we see that there is a period in which the subject of the crossed circle in Italy reaches a very great diffusion and therefore likely a particular importance: the Late Bronze Age (about 1150-950 BC). Indeed, this subject, even if it is recorded in rock art already in the Early Bronze Age, by the Final Bronze Age it is also present in many manufactured goods (Di Fraia 2004; Di Fraia in press).

It is known that the crossed circle is used in rock art, in various regions of the world, to represent the wheel, as evident, for example, in depictions of carts of the Andronovo culture (spread from Altai to the Urals in the second millennium BC) (Frankfurt 2003, Figure 1) or of the Iron Age Camuni culture (Anati 2007, Figure 31b). In turn, the wheel is used as a symbol of the sun (stylized round with rays) and this is evident in scenes where not only is it not part of a cart, but it is larger and clearly dominant over all other figures (Anati 1982, 109). We can assume that a further stage comprises those representations in which the sun appears by then completely independently, so that sometimes it is represented in a serial way.

As regards the possible meanings of the solar figure, Geoffroy de Saulieu (2004) suggested that in the Copper Age the sun—often associated with weapons and particularly with daggers—can be related to the fire of the forge, to the smelting of metal and to its gloss, with a sexual connotation concerning both the smiths (males) and the symbolic relationship between dagger and the masculine sex. In the monumental compositions (*art monumental*) "The sun could symbolize a hierarchy of which we can imagine that it was so visible as the sun, and so individual (but temporary) as is suggested by the ambiguity of the sun/human representation."

Later, during the Bronze Age, in more modest forms of art, especially in the sense that they were not intended towards ostentation (*art discret*) "to engrave a rock can be reduced to an individual or even private action, (de Saulieu 2004, 159) [translation of the writer].

If now we search some comparison to the crossed circle whether painted or engraved on the rock, we see that it does not find evidence to date in any central Italian rock art. Of course, we know that the argumentum *e silentio* in archaeology can never be a decisive discriminant, especially when one considers the extraordinary acceleration of new discoveries recorded just on the eastern slope of the Maiella during the last three years (Di Fraia and Manzi in press) .

It is then possible that new findings will reveal a wider distribution of this subject, in the same Abruzzi region or in neighbouring regions. Yet it is significant, in this author's opinion, the fact that among 44 rock art sites surveyed until 2007 (Mattioli 2007) in Abruzzi, Umbria and Lazio there is not one, to hold to the subject in question, that is not attested in Puglia either.

If the current record should reflect a really limited spread of this figure in parietal art of central Italy, we would be faced with an autonomous elaboration of patterns and symbols—however very common in the Late Bronze Age—that here are transposed into rock art forms by a population that must be rather small and isolated, at least with regard to religious imagery. Moreover, the acquired knowledge of rock art on the eastern side of the Maiella indicates an iconographic variety, argument in favor of a tendency to the creation of peculiar forms by individual human groups, although still we do not know how much possible chronological differences can count in this framework.

Going back to the Parete Manzi, if the crossed shield, the circle with the "x" and perhaps even the simple cross symbols represent the sun, as seems more than likely, we would be facing the presence of two fundamental entities: the feminine principle (Mother Goddess or Mother Earth, represented by the anthropomorphic imagery) with its values of fertility, reproduction and care, and the masculine (Sun) which combines power, energy, heat, light and life (Di Fraia 2004).

It is difficult, for the moment, to go over these suggestions with regard to religious concepts underlying these rock paintings. This will require still more research and studies.

References

ANATI, E. (1982). *Luine collina sacra.* Archivi, 8, Capo di Ponte, Edizioni del Centro.

ANATI, E. (2007). *Capire l'arte rupestre.* Studi camuni, vol. 26, Capo di Ponte, Edizioni del Centro.

CARABELLI, G. (1996). *In the image of Priapus.* London, Duckworth.

DE POMPEIS, V. (1993). Pitture rupestri in Abruzzo: nuove segnalazioni. *Atti della Società Toscana di Scienze Naturali*, 100, 65-83.

DE SAULIEU, G. (2004). *Art rupestre et statues-menhirs dans les Alpes. Des pierres et des pouvoirs (3000-2000 av. J. -Ch.).* Paris, Editions Errance.

DI FRAIA, T. (2004). Il motivo della ruota nei fornelli del Bronzo Finale. In N. NEGRONI CATACCHIO (ed.) Atti Sesto Incontro Studi di Preistoria e Protostoria in Etruria, 2002. Milano, Centro Studi di Preistoria e Archeologia, 235-244.

DI FRAIA, T. (2010). Immagini ornitomorfe e motivi solari nel Bronzo Recente e Finale in Italia. In P. Y. LE POGAM (ed.) *Imaginer et représenter l'au-delà*, 132e Congrès national des sociétés historiques et scientifiques, "Images et imagerie.", Arles, 16-21 avril 2007, Éditions CTHS, Paris, 7-22.

DI FRAIA, T. (2011). Prima della Grotta Di Ciccio Felice: il sito rupestre delle Pastine (Civitaluparella) e il rito dell'incubatio. In *Il Fucino e le aree limitrofe nell'antichità,* Atti del III Convegno di Archeologia, Avezzano 13-15 novembre 2009, 168-182

DI FRAIA, T. and MANZI, A. (in press). Nuove scoperte di arte rupestre in Abruzzo. In Atti XLII Riunione Scientifica dell'Istituto Italiano di Preistoria e protostoria, Trento, 2007.

FRANCFORT, H.-P. (2003). Les pétroglyphes d'Asie centrale: limites des interprétations indo-iraniennes et chamaniques. In J. GUILAINE (ed.) *Arts et symboles du Néolithique à la Protohistoire*. Éditions Errance, Paris, 189-214.

FUCINESE, D. V. (1995). *L'eremo di San Venanzio a Raiano*. Pescara, Edizioni Tracce.

GROSSI, G. 2000. *Le origini di Avezzano*. Avezzano, Rotary Club.

MANZI, G. and MANZI, A. (2004). *Permanenze ed evoluzione della cultura italica nell'Alto vastese*. Lanciano (CH), Edizione Alessandro Lanci - Allestimenti museali.

MANZI, G. and MANZI, A. (2007). *Pastori, lanaioli e contadini. La pastorizia e la lavorazione della lana nel versante orientale della Maiella.* Sambuceto (CH), META Edizioni.

MATTIOLI, T. (2007). L'arte rupestre in Italia centrale. Umbria, Lazio, Abruzzo. Quaderni di Protostoria, 4. Perugia, Ali§no.

MICATI, E. (1990). *Eremi e luoghi di culto rupestri della Majella e del Morrone*. Pescara, Carsa Edizioni.

MICATI, E. (2002), *Gli eremi della montagna teramana.* Teramo, Edigrafital.

MOSTACCI, D. (1997). Pitture rupestri a Raiano (L'Aquila). *Bullettino del Centro Camuno di Studi Preistorici*, XXX, 66.

PANSA, G. (1924). *Miti, leggende e superstizioni dell'Abruzzo*. Ristampa anastatica a cura di Arnaldo Forni, Bologna, 1981.

RADMILLI, A. M. (1956). Preistoria e protostoria marsicana: gli scavi nella Grotta "Di Ciccio Felice". *Rivista di Scienze Preistoriche*, XI, 31-52.

The Mother Goddess in the Monuments and Rituals at the "Dausdava" Getic Religious Centre: Thracian Remains and Modern Relicts

Diana Gergova

National Archaeological Institute with Museum, Sofia
dianagergova@abv.bg

Abstract*. The intensive interdisciplinary investigations of the Getic religious and political centre of Dausdava, dated to the 1st millennium BC, resulted in the discovery in the last 25 years of diverse monuments—tombs and sanctuaries, altars and reliefs, images, inscriptions and traces of rituals. They reveal the main role of the cult to the Mother Goddess among the Northern Thracians—the Getae. The tradition of the sacred place has been preserved even now. The continuity of the rituals and beliefs of ancient Indo-European origin, performed at the places of the ancient Thracian sanctuaries by the local population, underlines the extreme strength of the cult to the Mother Goddess that has been preserved despite the ethnical and religious changes that took place through the millennia. These traditions include a belief in the principle of 'incubatio' and the use of holed stones.*

Key words: *Dausdava, Getae, Holed Stone, Incubatio, Mother Goddess, traditions.*

Introduction

The Sboryanovo National Reserve strikes the investigator with the richness of data and the variety of expressions of the veneration of the Mother Goddess, from the foundation of the Getic religious and political center at the beginning of the Iron Age and until modern times she is eternal.

Intensive interdisciplinary investigations of the Getic centre—most probably Dausdava, (The City of the Wolves, or The City of the Dawn) on Tabula Nona of Claudius Ptolemeus, dated to the first millennium BC—as well as of the later monuments on its territory started with the discovery of the Sveshtari Getic royal tomb in 1982. Its uniqueness gave impetus to start investigations of all the main elements of the Getic capital in order to understand the cultural and natural background of the unique tomb. Sanctuaries, images, inscriptions from the first millennium BC, as well as later monuments and traces of rituals that have been preserved till modern times, are proof for the eternity of the cult to the Mother goddess, whose strongest presence is in the first of the discovered Getic tombs—the Sveshtari tomb. (Fol, Chichikova, Ivanov, Teofilov 1986, 64) (Fig. 1).

Descriptions

The interior of the Sveshtari tomb—dated to the end of the fourth and beginning of the third century BC—and especially of its main chamber resemble the nave of a temple: a temple of the Mother Goddess. The suggestion is supported by the images of the 10 caryatids in high relief. They are chiseled in high relief on panels of soft yellow-white limestone. There are three figures on the north-eastern and south-western walls and four on the north-western wall. They wear long sleeveless dresses (chytons), girded with a narrow belt below the breasts, with a deep apoptygma. The chyton is rendered decoratively as an upturned calyx made of three acanthus leaves. The recurrent gesture of the arms raised as if in prayer produces a peculiar rhythm to the composition. The heads of the caryatids are topped by calathi. The faces are individual. Mature women are featured along with young idealized faces. (Fol, Chichikova, Ivanov, Teofilov 1986, 64)

The motif of the female deity combined with floral ornaments was widespread in antiquity and is associated with the image of the Great Mother Goddess, as Potnia phtion in the Mediterranean world in the Bronze Age, and reappear in almost all countries of the same area at the end of the fifth and fourth century BC. The Hellenistic iconography of the Great Mother is well known from the Northern Black Sea area, Macedonia, South Italy (Fol et al. 1986, 116-117). It is important to note that this iconography of the Goddess of life and death, of fertility and birth, uniting the earth and the sky becomes especially characteristic for some Mediterranean area that are well known for their very close connection with the Orphic traditions (Gergova 1996; Gergova 2007b).

The individuality of the faces of the women gives reason to suggest that they might be portraits of real women. It has also been suggested that the ten women were fulfilling a ritual dance. Amazing is the contrast between the realism of the women's faces and their raised hands—stylized and imitating tree branches. Thus the ten caryatids combine

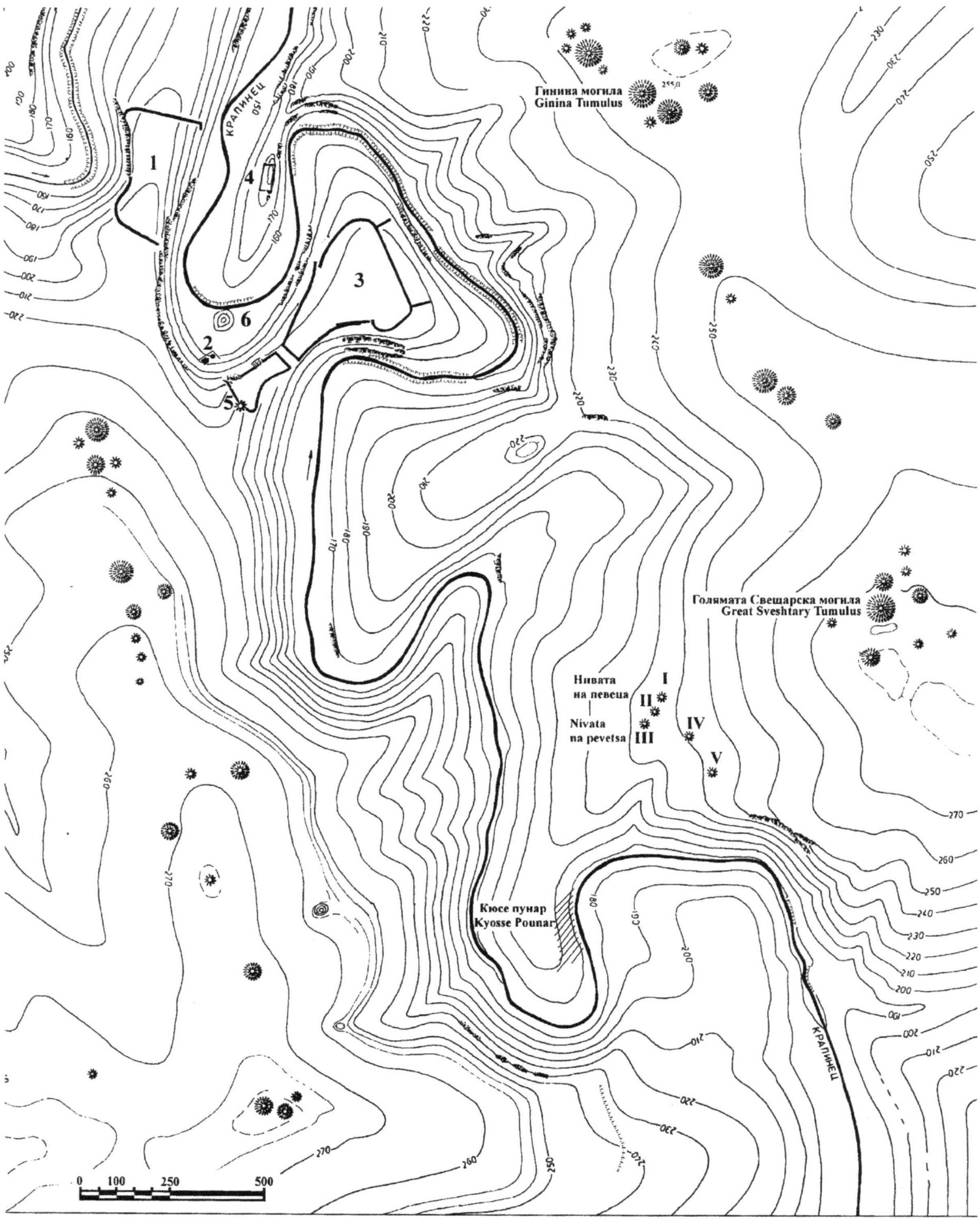

1 Камен рид	4 Късноантичен кастел	1 Kamen rid	4 Late Antiquity Fort
2 Демир баба теке	5 Ранновизантийска крепост	2 Demir Baba Teke	5 Early Byzantine Fortress
3 Тракийски град при Водната централа	Селище от рнно желязната епоха	3 Thracian city at the Water Supplying Station	Early Iron Age Settlement
	6. Праисторическа селищна могила		

Fig. 1. Plan of the Sboryanovo reserve

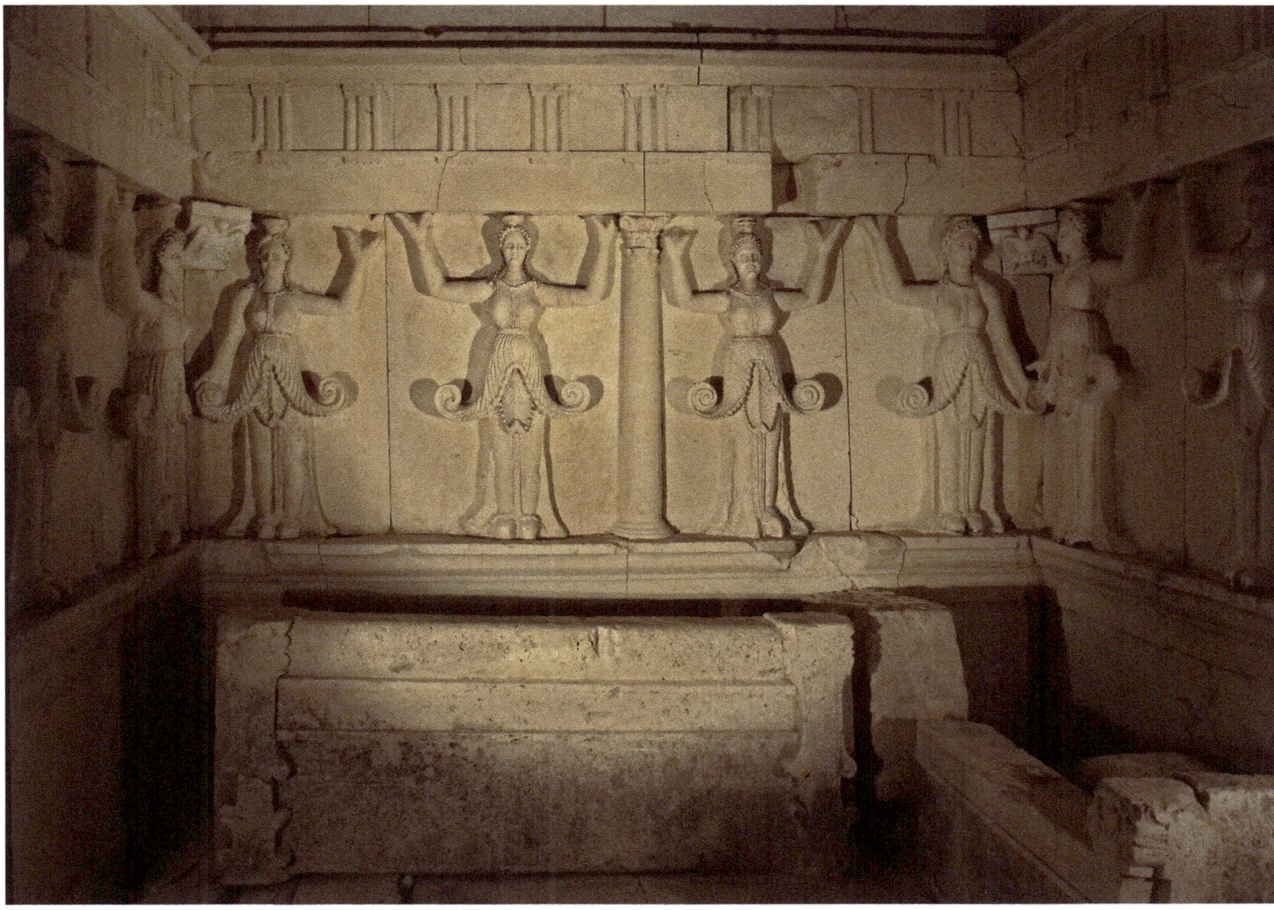

Fig. 2. The main chamber of the Sveshtari tomb

realistic faces of women or different aspects of the image of the Goddess with the symbolism of her iconography, underlining at the same time the specific connections with the Tree of Life as one of her main symbols (Fig. 2, 3, 4).

The Mother goddess is the main figure in the scene of the hero-making of the king represented on the NW lunette of the tomb. It depicts the horseman, who is actually the deceased receiving the wreath of the immortalization from the Mother Goddess to become Hero. She stands on a podium and her figure is larger than those of the other participants in the ceremony—warriors and servants (Fol et al. 1986, 74) (Fig.5). This scene according to some archaeo-astronomic studies was reached by sunlight, penetrating a small opening in the tomb exactly on the day of the winter solstice at the end of the fourth century BC. Thus the whole cycle of death and rebirth was part of the cosmic harmony (Гергова, Кадийска 2003)

The excavations of the several groups of tumuli of the Hellenistic necropolis of the Getae, planned and piled as mirror reflections of some of the brighter constellations (Canis major, Orion) gave more data about specific aspects of the burials of the Getae, hinting also at a special place for the Mother Goddess in their beliefs and burial practices. A special type of dug-out in the soil tombs, sometimes strengthened by additional wooden structure and roof, almost always have a rectangular clay altar in the centre

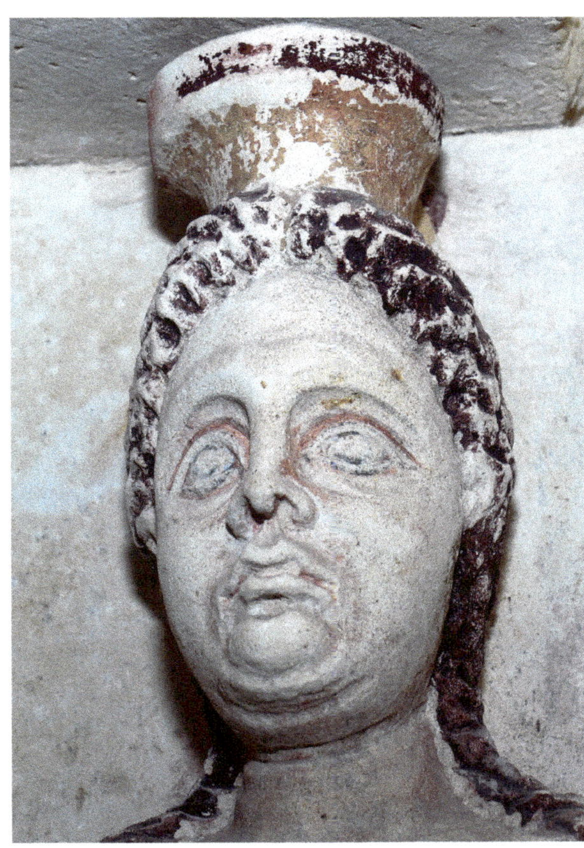

Fig. 3. The face of one of the caryatids

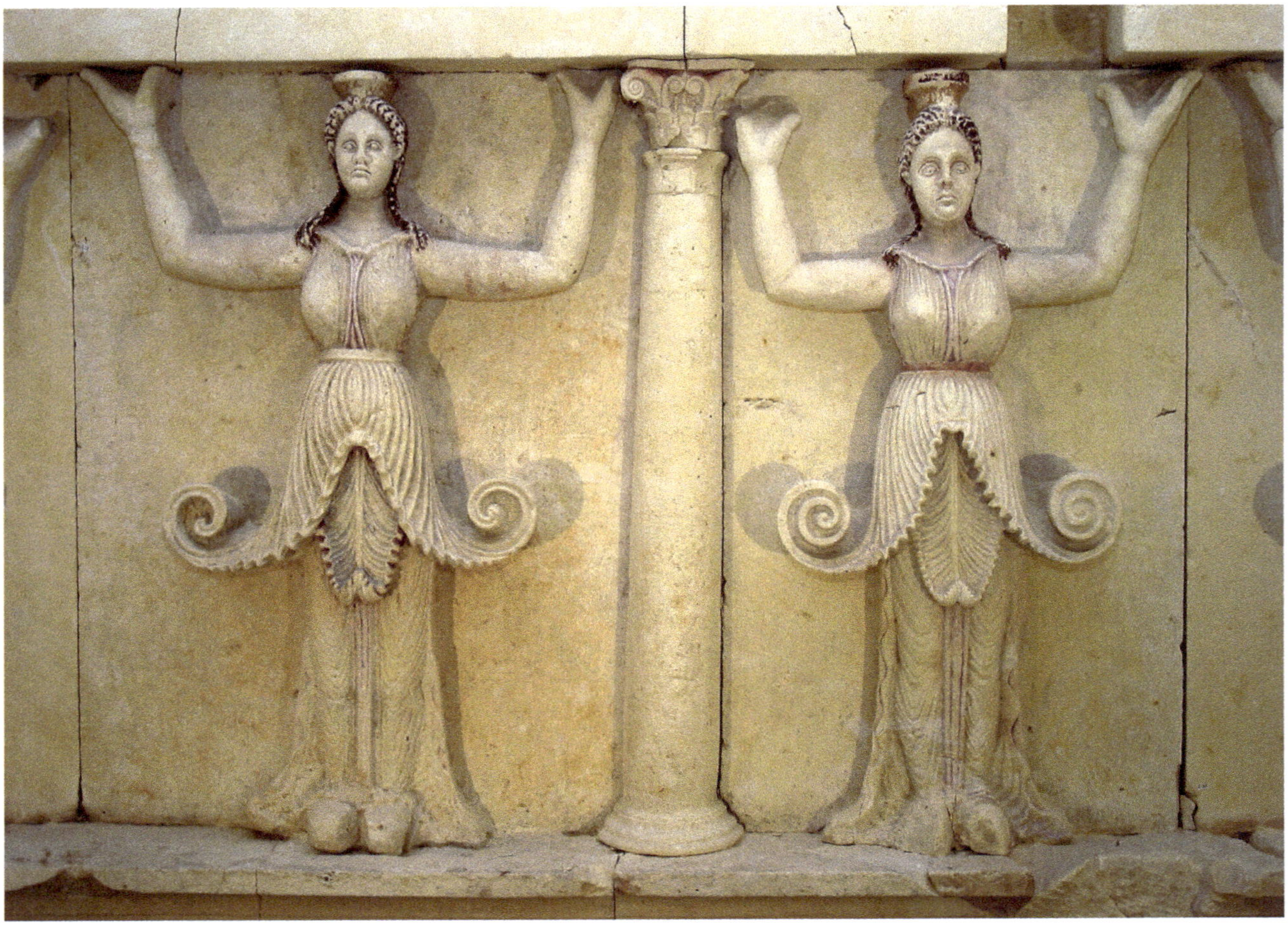

Fig. 4. Two caryatids

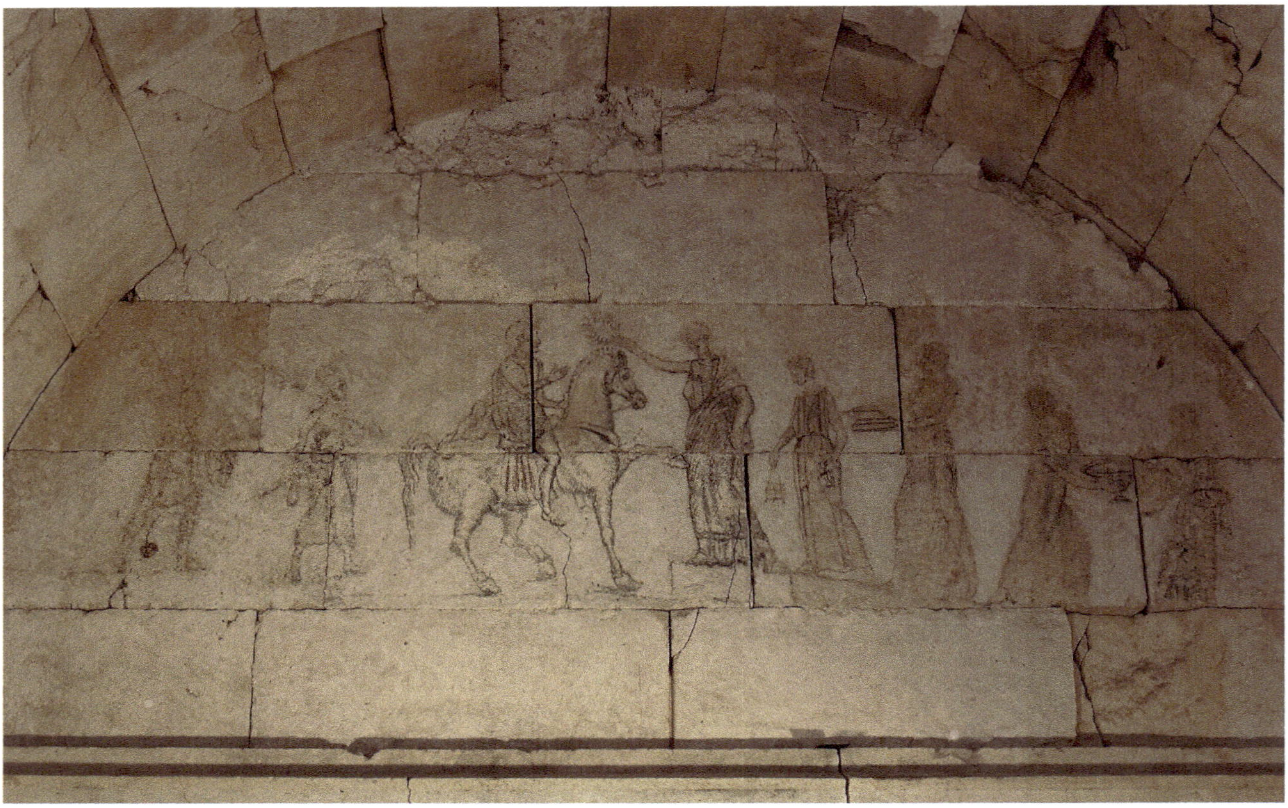

Fig. 5. The scene of the immortalization

of their main chamber. Several escahae have been found in tombs on the territory of the Eastern, royal necropolis of the Getic capital (tumuli 18, 21, 4) (Гергова 1991; Гергова, Катевски, Иванов, Димчева, 2005, 19-21, 44-50; Гергова, Иванов 2011; Гергова, Радев 1991, 46-47) and one portable in tumulus 1 of the Western necropolis (Гергова, Кузманов 2006). The inventory of these tombs suggests priestly or priestly-warrior functions of the deceased.

The most interesting is the first one, discovered by G. Feher in one of the tumuli of the Southern group, near the highest tumulus of the necropolis—the Great Sveshtari tumulus. The incised geometric ornamentation includes four "M" signs. Vl. Georgiev considers that in the Thracian language M would correspond to Mother (Георгиев 1977, 86) (Fig. 6), a suggestion that becomes even more important when having in mind the cults with which the escharae had been connected (Чичикова 1975; Gergova 2006).

The origin of the escharae can be traced back to the Late Bronze Age Mediterranean and SE European world. They are elements of the sacred architecture and of the ritual practices especially strong on the island of Crete and also known from the Thracian sanctuaries in the Rhodopi mountain of Tatul and Babyak, as well as from Romania. The escharae reappear similarly to the image of the Mother Goddess and become extremely popular in Thrace in the Hellenistic period. They are also known from Southern France and Northern Europe and it has been suggested that the processes were interrelated. (Makiewicz 1987; Marty, del Corso 2002; Gergova 2006**).** In Thrace escharae appear in temenoi and temples, from palaces and houses like Seuthopols, Helis, Emporion Pistiros, Kabyle, Philipopolis, etc. (Чичикова 1975; Михайлова 2005; Dimitrova - Milceva 1990,;Lazov 1996; Гергова 2008). The decoration of the escharae with snakes, floral or geometric motives indicate once again their connection with the Cretan Goddess with the snakes that are not characteristic for the Greek, but for the non-Greek religious traditions. According to the Seuthopolos inscription the clay altar in the palace in Sethopolis was connected with the cult to the Great Gods of Samothrace (Чичикова 1975,Gergova 2006). It is important to note that among these gods the main role is played again by the Mother Goddess.

The image of the Getae as a tribe closely connected with the ideas of the immortality and Pythagoreanism and their consequent propagation, is thus approved by the construction of the escharae as specific religious symbols in their graves. Thus we can see that Thrace and in particular the Getic lands have preserved the ancient religious traditions of the Eastern Mediterranean world in which the Mother Goddess plays the main role in the best way.

It is worth mentioning the only inscription, found till now on the territory of Dausdava, or more precisely near the Southern gate of the Hellenistic town and capital of the Getae. Most probably this is the polis of Helis, mentioned by Diodorus in connection with the victory of the Getic

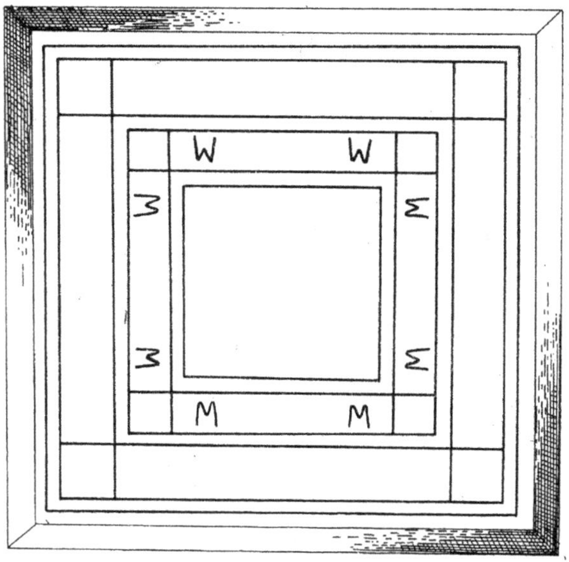

Обр. 1. Есхарата от проучванията на Г. Фехер – източен некропол на Сборяново

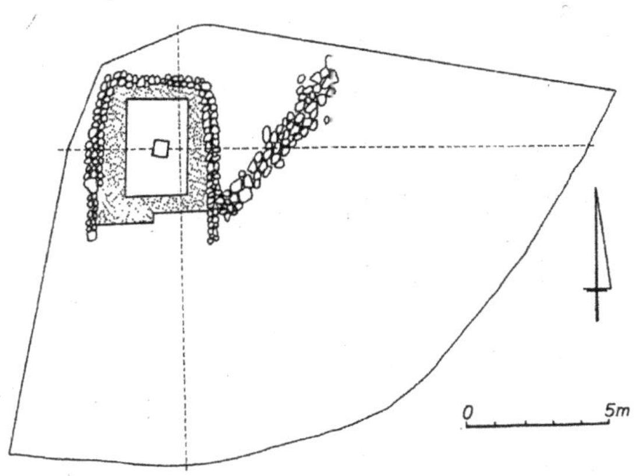

Fig. 6. The eschara with the letter M

king Dromichates over Lysimachus, and the reception that the Getic king organized there (Diod…21,11;21,12,3). The inscription on a lime block is a dedication from the Greek Posidonius, son of Menecharmos to Phosphorus (Artemis) Phosphorus was venerated as a patron of the towns, and usually small temples were built near the city gates, similar to that of Helis (Fig.7)) (Чичикова 1990; Gergova 2004, 44).

The Hellenistic images, sacred structures and inscriptions reveal the main place that the Mother Goddess occupied in the road to the afterlife and the mysteries, as well as in the everyday life. Our investigations revealed that the beginning of these beliefs and the cult of the Mother goddess had earlier expression in the Getic lands, in the beginning of the Early Iron age, when the Getae founded their sacred centre near the only water spring in the area, in the picturesque valley of the river Krapinec. There they founded their two sanctuaries on the plateau Kamen Rid and in the valley, around the rocks and the water spring,

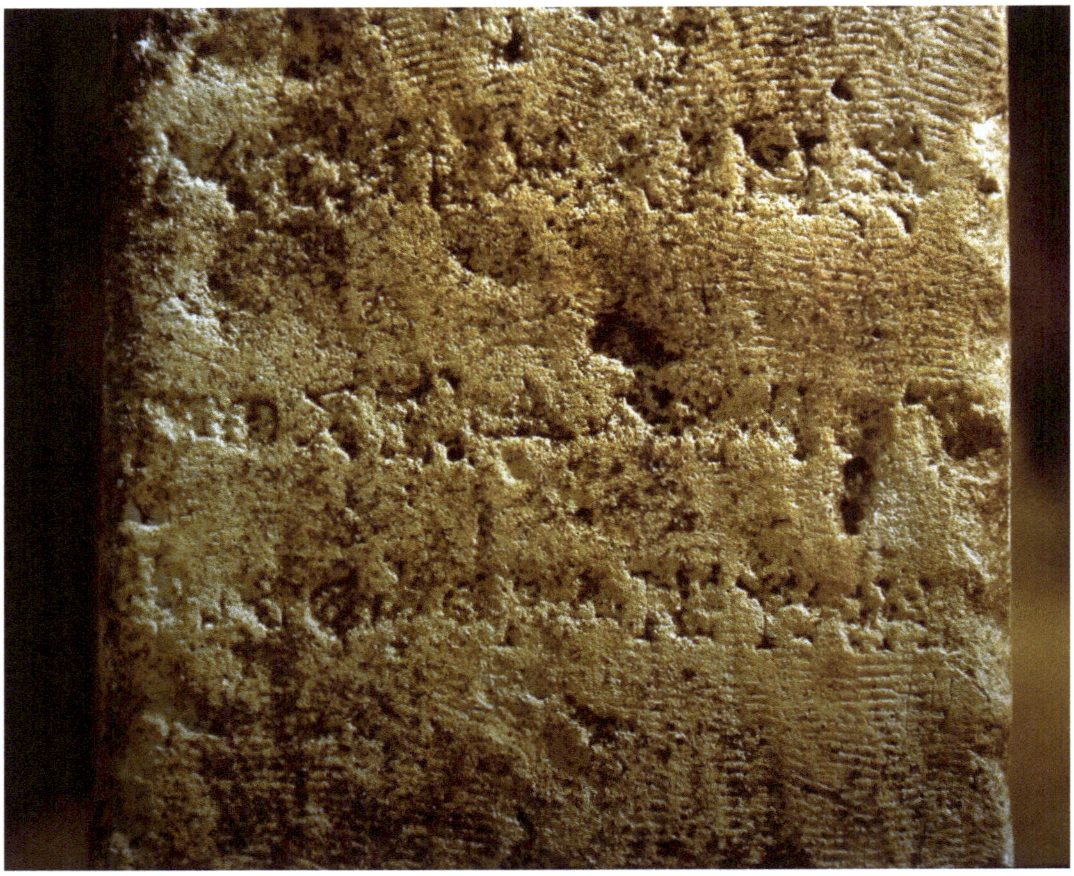

Fig. 7. The inscription dedicated to Phosphorus

called later "The five fingers" (Balkanska 1998; Gergova 2004; Gergova 2007a.)

The temenos at Kamen Rid covers an area of 47,000 m^2 and is surrounded by stone walls. The earliest dating to the Early Iron Age—eleventh to tenth century BC—surrounded specific cult structures like fireplaces and ritual pits, usually connected with the chthonic gods. Clay female figurines, characteristic for this period and known from many other cult places and burials, originate from the pits. There are also male figurines, indicating that since the beginning of the Iron Age we will very often see the specific relationships between the woman—the Mother Goddess—and the man, her son Apollo—or the horseman and king—to become immortal if blessed by her (Fig. 6) (Gergova 1988a; 2007a).

A renovation of the sanctuary took place in the Hellenistic period. New stone walls following the direction of the earlier ones were built and a system of two circles along the axis S-N was built in the Northern part of the temenos. The Northern circle has an entrance from the NE and a tree growing in its centre. The southern circle has a big rock fragment incorporated in its wall and other semicircles around it. Although there are no more precise data for the attribution of the sanctuary to certain god or gods, it seems quite probable that the two circles might be a materialization of the idea, connected with the celebration of the Sacred Triad of the Hyperborean deities: the Mother Goddess and her twins, Apollo and Artemis. In Delos, where Leto delivered her twins, the triad was celebrated by a dance around the sacred tree of Artemis and by another one around Apollo's altar (Gergova 2004; 2007a; Гергова, Теодоров 2006).

The earlier, Early Iron Age clay female and male figurines also suggest the connection with the main gods, venerated by the Thracians, according to written tradition. It is important to note that similar figurines were found in the Early Iron Age necropolis in the "Singer's field" locality belonging most probably to the Getae, connected with the maintenance of the temenos. Here in one of the tumuli a female burial of a priestess was found accompanied by a clay figurine (Stoyanov 1992, 141-143, 151).

The sanctuary on Kamen Rid had, and has even now, a direct connection by a narrow path along the rocks and caves with the other sanctuary founded by the Getae: the valley of the river with "The five fingers", the water spring. This place is the core of the sacred Getic area (Fig.1). The rocks of the canyon of the river are forming naturally figures of people and animals. Thus one of the rock protuberances, as seen from the plateau is anthropomorphized and named "The head of Demir Baba" (Fig.9). Another, controversial for its human or natural origin as the figure of a horse or a dog, can be seen from the courtyard of the teke (Fig.10). Most probably, this sanctuary was also founded in the beginning of the Early Iron age. A megalithic rock step altar, similar to the earliest altars in Delphi and Western

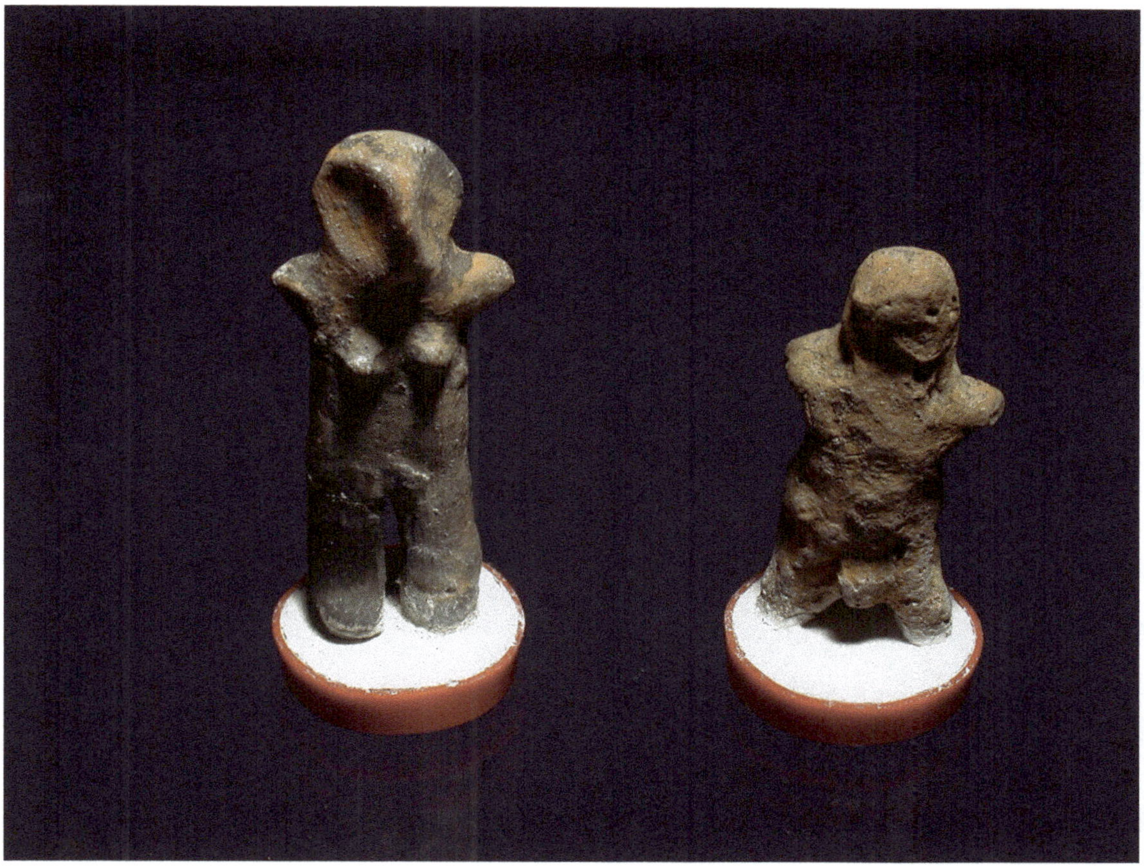

Fig. 8. Clay figurines from Kamen Rid sanctuary

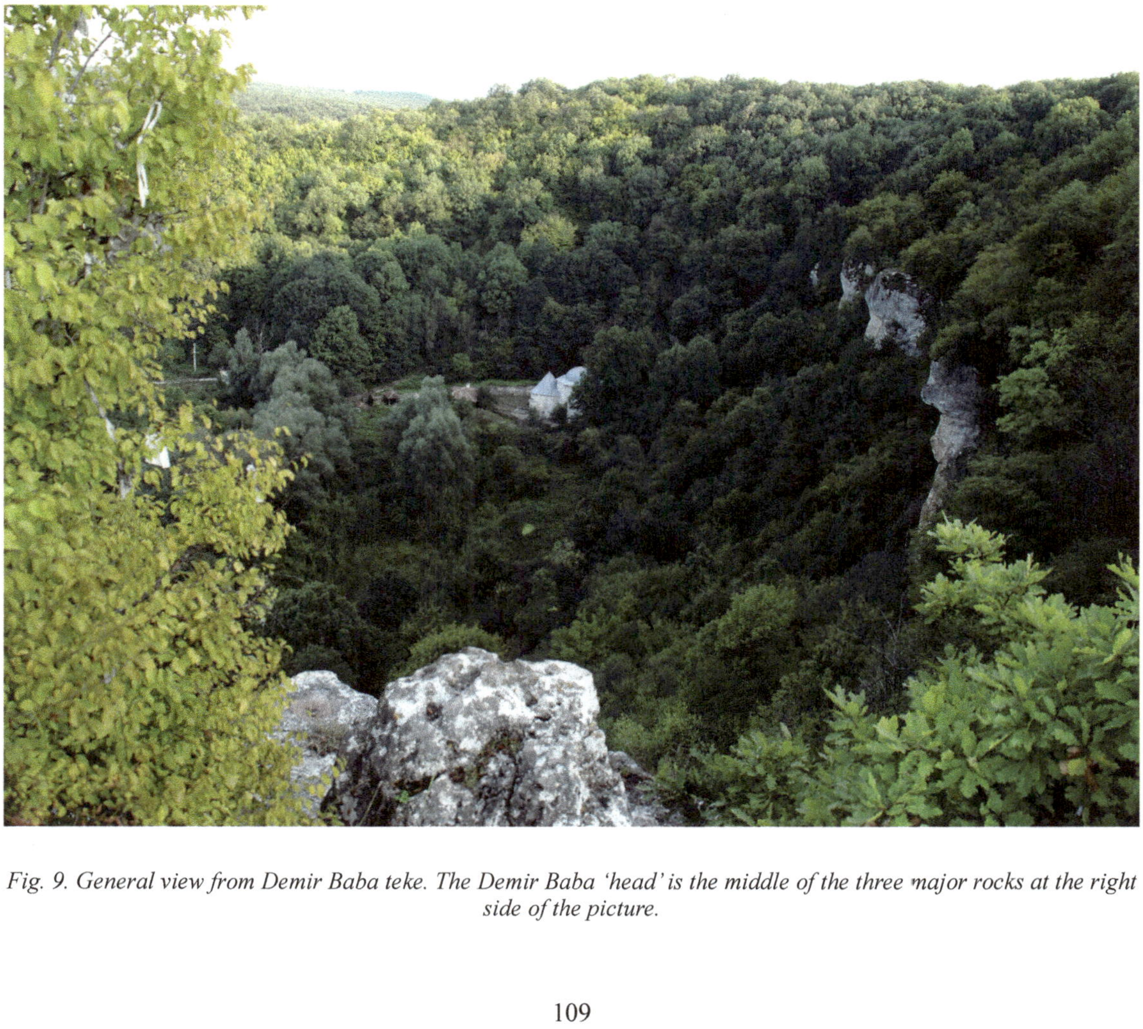

Fig. 9. General view from Demir Baba teke. The Demir Baba 'head' is the middle of the three major rocks at the right side of the picture.

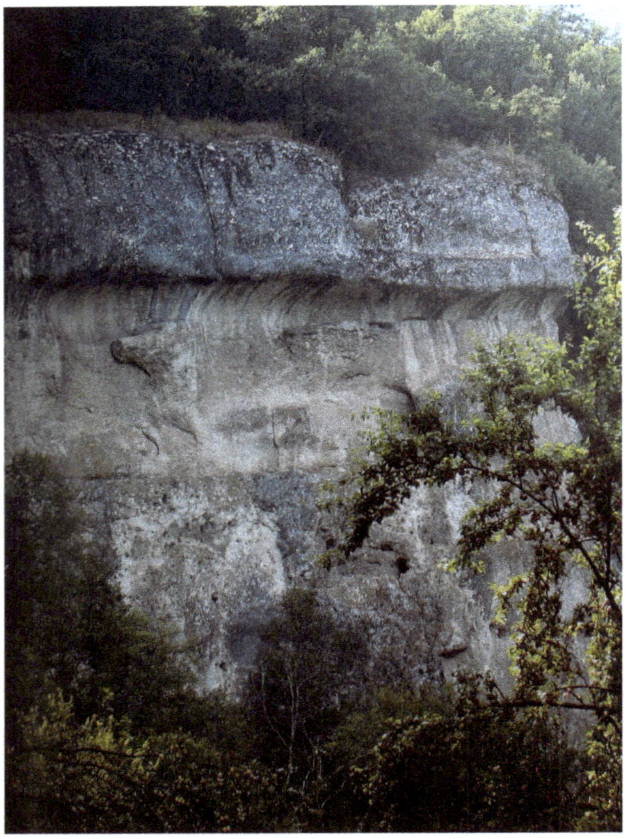

Fig. 10. A view of the 'horse' in the rocks as seen from the direction of Demir Baba teke.

Asia Minor from the end of the Bronze and the beginning of the Early Iron Age can be seen now incorporated in the Late Medieval teke. Another one, with grooves for the blood of the sacrificed animals has been preserved till now and included in the stone gate to the courtyard of the teke. Similar to the temenos on Kamen Rid, this sanctuary was reconstructed in the Hellenistic period, when the Getic centre became the political center of the Getae. Then a small temple with two omphaloi—Apollo's symbols—and two ritual tables was built. Many clay altars were found around it, under the floor of the teke (Balkanska 1998). Later excavations in 2004 discovered another omphalos in the same area, most probably from an earlier phase of the temple (Гергова 2007; Гергова, Венедикова 2007, 35).

This sanctuary continued to function in the Roman period as a sanctuary of the Thracian horseman (Gergova 2004, 53,57; Гергова, Венедикова 2007,45). Its later history is connected both with Christianity as far as Karel Shkorpil saw when visiting the site at the end of the nineteenth century (a stone cross on the hill near the former pagan sanctuary) as well as with the Alevian monastery of Demir Baba (Гергова, Венедикова 2007,46)

Before discussing the later preserved sacred architectural monuments dated to the Late Medieval period, it is helpful to say several words about the Alevians, called also Kazilbashi—a confessional group inhabiting the villages in the neighbourhood and also known from other parts of Bulgaria. The Alevian teke of Demir Baba was founded by people coming from Iran, Horasan, Azerbaijan and Volga river area confessing Islam, but connected with the idea of the common roots of the different religions and with the propagation of the idea about equality of people under the sun. The role of the woman was very important in the religious life of the Kazilbashi, and it was obligatory that ritual gatherings were attended by the couples. The beliefs of the followers of Ali, the son-in-law of Mohammed, were strongly influenced by Zoroastrian, Dionysian and Pythagorean ideas. It seems that this was also one of the reasons to build the teke in the fifteenth century in the valley of the river, on the megalithic altars of the earlier Thracian sanctuary, instead of following the principle to build the teke on high hills (Гергова, Венедикова 2007).

The teke incorporated or transformed some of the earlier Thracian cult structures for the need of the new sacral architecture and rituals. It was below the floor of the teke that most of the clay Hellenistic fireplaces were found during the excavations (Balkanska 1998). The Thracian step altar was preserved in the Northern wall of the teke, while other megalithic elements were cut for the fulfillment of some rituals.

The decoration of the teke—in fact a tomb-temple of the philosopher and healer, warrior and hero of Bulgarian origin, the mythological Demir Baba—is hinting at the special, although hidden role of woman in the beliefs of the Kazilbashi. Extremely interesting inscriptions—decorations—were discovered on the dome of the antechamber during its restoration about 20 years ago (Fig. 11). The first of them contains graphic signs with the names of Allah, Mohammed and Ali, a sacred triad similar to the Christian one. The other three identical inscriptions are hiding behind their calligraphy the images of three stylized maidens or women's smiling faces. They are similar, but painted in different colours—light blue, green, and light blue again. They are formed by the symmetrical presentation of the letter "vav" that resembles a comma. The eyes are elliptical. The nose and lips are also marked. The ears and hair are represented in a different way. The motifs are crowned with different tips: one with a tulip, the second with a miniature stylized bough, and the third with an image reminiscent of zulfikars' sword The motifs are beautifully written letters containing an esoteric meaning and connected with a highly heretical trend in Shiite Islam, Hurufya. It appeared in Iran or Azerbaijan in the late fourteenth century and was of a markedly anti-Islamic character. Pythagoreanism played a key role in the philosophy of Hurufism. Numbers were fraught with symbolism, while letters were mystical signs of the lines and elements of the human face. The main idea of the teaching was that God is in man. Such persecuted teaching penetrated also among the Kazilbashis (Gergova 2004; Гергова, Венедикова 2007).

The symmetrical "vav" was symbol of the 38th enichar's

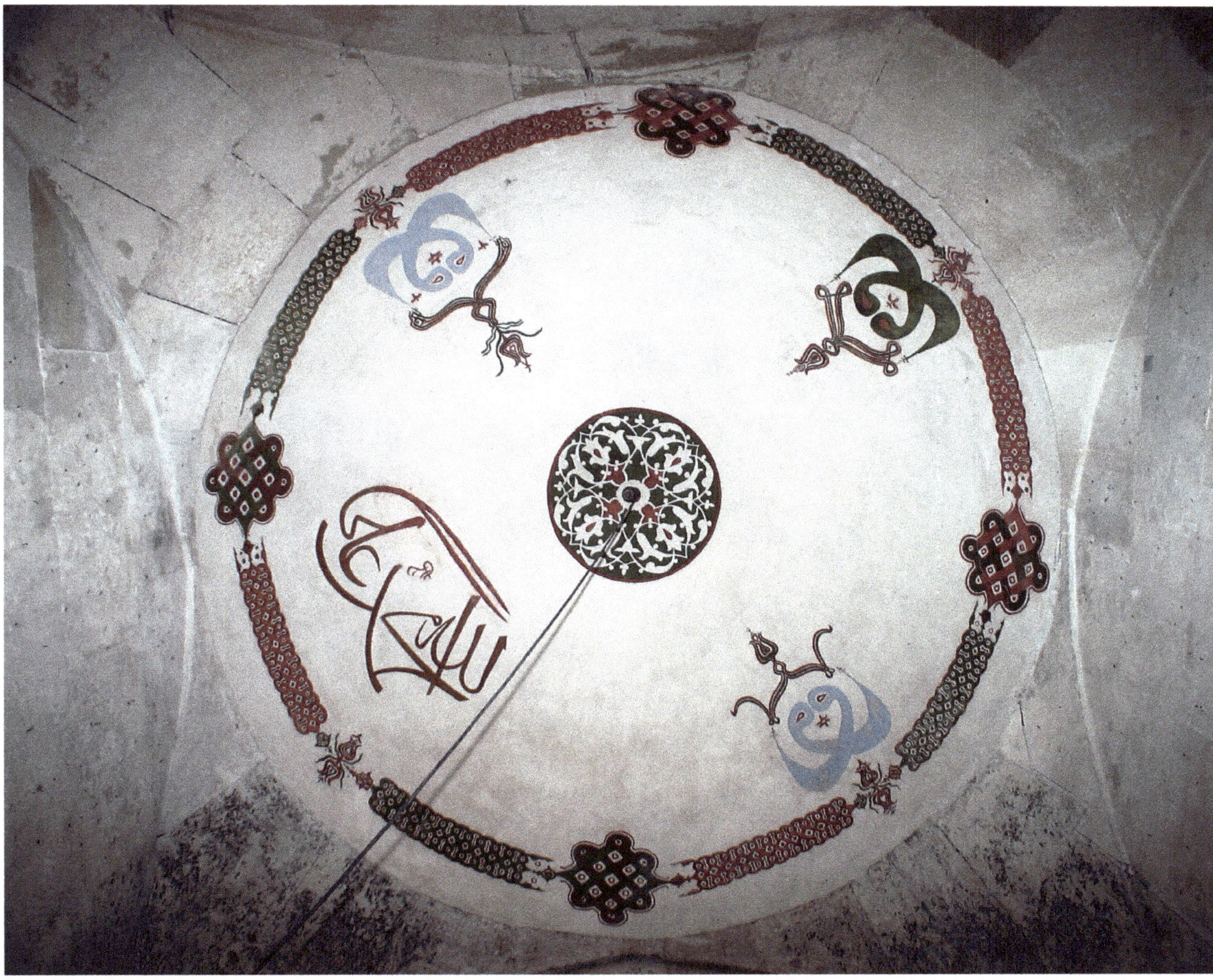

Fig. 11. The decorated vault of the antechamber of the teke

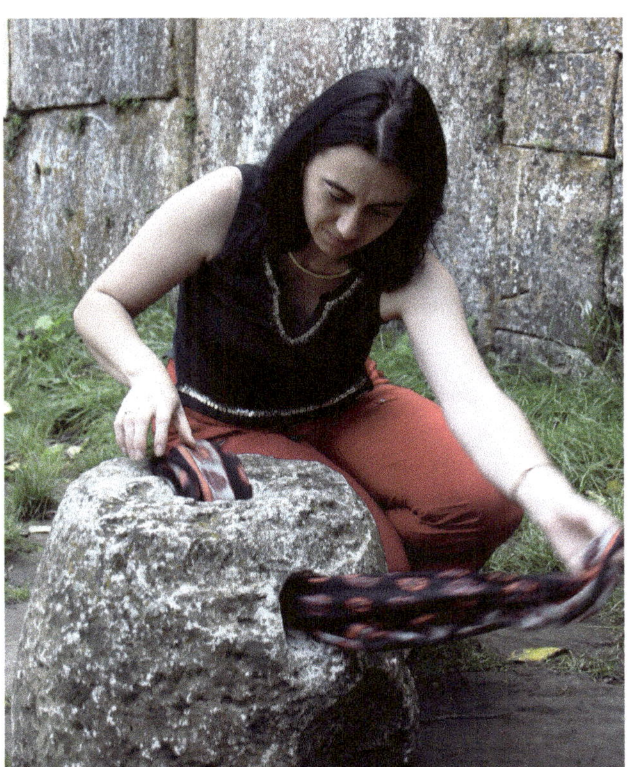

division (Гергова, Венедикова 2007). Some authors, having in mind the ideas of Hurufism are trying to connect the female faces even with the symbolic presence of Fatima—the daughter of Mohammed and wife of Ali (Овчаров 2006).

The building of the sanctuary over the megalithic Thracian altar, the huge stone blocks built into the construction and into the stone wall, suggest a special aesthetic quality of the megalithic culture, as well as a cult of the rock, fire, water and Nature. The interesting and mysterious customs of the Alians—that survived in their closed society because of the persecutions of the official Ottoman authorities—performed in the courtyard in modern times are of enormous interest. In the court of the teke you can see women to passing their underwear and scarves through the holes of an ancient capital to take the strength of the rock for health and fertility (Fig.12) To climb to the top of the Thracian step altar is another tradition in the courtyard of Demir Baba teke (Gergova 2004).

Fig. 12. A long-lived tradition in the court yard: Passing a garment through a hole in a stone.

The incline to the East Rock in the southern wall of the teke was cut in such a way that people could lie on the sloping stone (Balkanska1998), their head to the east—best before noon when the sun is in the east so they can draw upon the force of the Earth—the Mother Goddess (Gergova 2004) (Fig. 13). We can certainly interpret the ritual as a symbolic "incubation"—the religious practice of sleeping in a sacred area with the intention of experiencing a divinely inspired dream or cure. A good European example of incubation is that reported by Tomaso di Fraia for the cult site Civitaluparella in Italy (Di Fraia 2012). Prof. Meaden has proposed that something similar (akin to an experience of the sacred) could explain the purpose of the single body-length recumbent stone found at several stone circles in Scotland and Ireland (Meaden 1991, 184-185). Some have a recognizable astronomical orientation and are aligned to the midwinter sunset or midsummer sunrise (as at Drombeg in Cork) or to moonsets. I am very grateful to Prof. Meaden for his suggestions and personal information, because he adds more details to the question of the specific Northern, Hyperborean traditions, that find their materialization in the ritual practices of ancient Thrace, and link South-Eastern and North.

It is important to note that the incubation at Demir Baba teke is a common practice both for Christians and Muslims. As a religious practice of Islam it is called *Salaat-ul-Istikhaarah*. The incubation is the intention of experiencing and searching for good things, asking Allah to help one make a choice, choosing the best of two things where one needs to decide on one of them (Wikipedia). During his visit to Sboryanovo and Demir Baba teke several years ago, Prof. T. Hayashi from Japan shared his impressions from a visit to Iran, where the same tradition is widely spread, but with a difference that underlines the connection of this rite with the Mother Goddess. According to him only women have visited these types of sanctuary and the same rituals have been performed only by them.

Adorning trees with items of clothing and coloured ribbons, as done predominantly by young girls, is particularly interesting. This is essentially the symbolic clothing of the Great Goddess, or her Tree of Life. The tradition is well known from the Balkans, Asia Minor, and Asia, especially in Ephesus with the famous temple of Artemis and the grave of the Virgin Mary. Usually connected with later Islamic traditions, the performance of the same rituals in Northern Europe—the Isle of Man, in sacred places in Scotland and Ireland, in the forests of Denmark and northern France—is proof now for the ancient, Indo-European root of the tradition whose ramifications permeate all who believe in the divine nature of mankind (Fig. 14) (Gergova 2004).

What is even more important is the origin of the word teke, considered for a long time to be of Turkic origin. It turned out that teke was a word that could be found in the Ancient Greek vocabulary and means " a tomb"," a box". This was the word used for the graves of the mythical Hyperborean maidens in Delos-Hyperoche and Laodice (Гергова, Венедикова 2007, 89). Thus we have another example of the strength of the ancient Mediterranean traditions preserved in the name of the tomb of Demir Baba.

Concluding remarks

The Sboryanovo reserve—a unique sacred place—reveals in a convincing way the deepness and eternity of the veneration of Nature and the Mother Goddess. The Getic religious and political center of Dausdava was founded at the end of the first millennium BC in a mystic natural milieu where Nature had shaped fantastic anthropomorphic and zoomorphic images during the millennia, and offered most favorable conditions both for life and meditation. Planned according to the principles of the Orphic doctrine of immortality propagated by the king, the priest and the god of the Getae Zalmoxis, the Getic sacred area has preserved until now monuments that illustrate the central position of the cult to the Mother Goddess, expressed by sculptures, altars, texts and paintings. These Orphic and Pythagorean traditions of the Getae remained vital, and influenced with their universalism the later religions of similar esoteric character. In the belief of the immortality of the soul, in the presence of god in man, in the sanctity of Nature in which the Earth, the female principle and the Mother Goddess always played the main role.

References

Георгиев Вл. (1977). Траките и техният език.

Гергова, Д. (1987). Съкровищата от Тракия и хиперборейския мит. – В: Българските земи в древността. България през средновековието. Доклади на Втория международен конгрес по българистика. Т. 6, 1987, 53–74.

Гергова, Д. (2002). Елинистическият некропол на Сборяново.- В: По стъпалата на времето. Каталог, 30-40, София

Гергова, Д. (2002). Източен могилен некропол.-В: По стъпалата на времето. Каталог, 64-66, София

Гергова Д. (2007). Нови данни за сакралната архитектура в двора на Демир баба теке .Изследвания в памет на акад. Анание Явашов. Академична издателство „Марин Дринов", София, 47–53.

Гергова Д. (2008). Есхарата в гетското погребение. – In: *PHOSPHORION. Studia in honorem Mariae Cicikova*. Sofia, 255–267.

Гергова, Д., К. Венедикова (2007). Демир баба теке. Българският Ерусалим. Агато, 112 с.

Гергова Д., Иванов Я., Катевски Ил. (2010). Проучванията на елинистическия некропол на гетския култов и политически център в Сборяново.—АОР за 2010, Софияр 2011, 205-207

Гергова, Д., Т. Кадийска (2003). Светлината в сакралната архитектура на тракийските гети. – Детайли, /№ 1, 6–10.

Гергова Д., Катевски Ил, Иванов Я., Димчева Д. (2005). Могила 18. "Сборяново" Еленистическият некропол (проучвания през 2004 г.). Getica. кн.1.

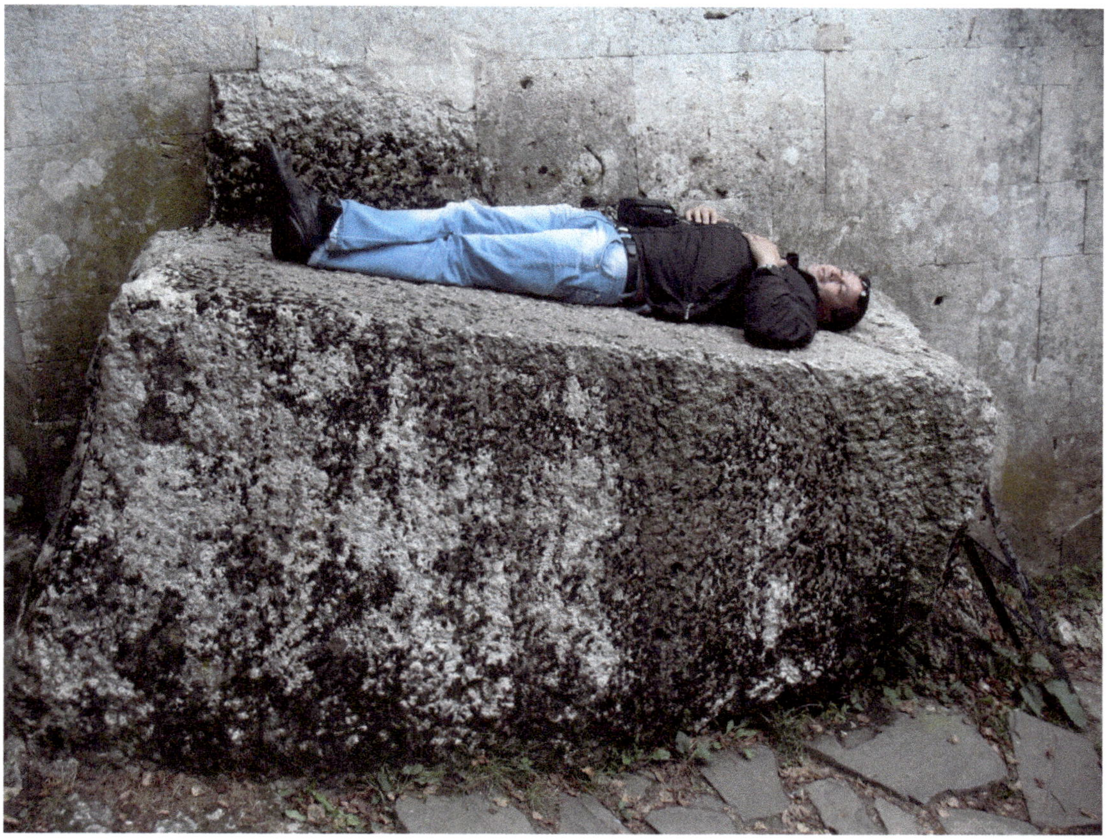

Fig. 13. Another tradition in the courtyard: Lying on a sacred megalith.

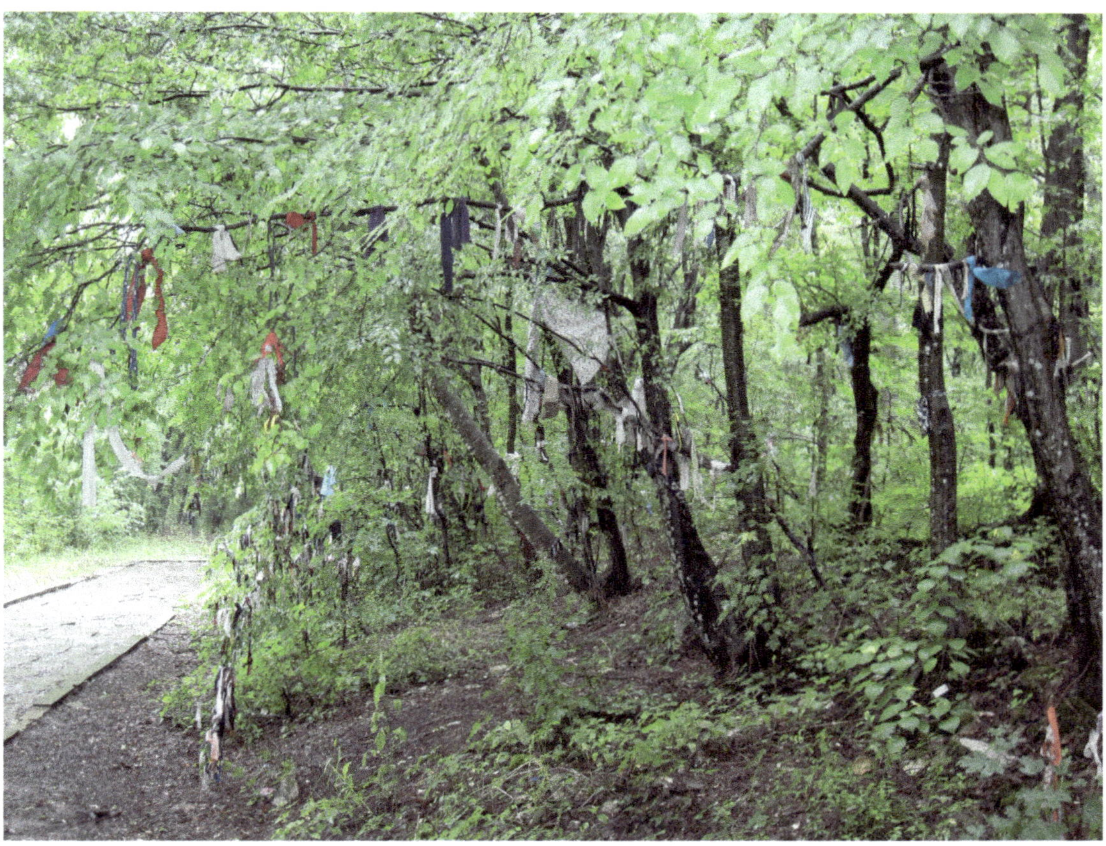

Fig. 14.Trees of Life hung with ribbons.

Гергова, Д., М.. Кузманов (2006). Проучвания на Могила 1 от западния елинистически некропол на Сборяново. – **В:** Хелис, V, 212–228..

Гергова, Д., Е. Теодоров (2006). Прабългарски и тракийски следи. Изд. "Изток-Запад", 124 с.

Гергова, Д., Радев Р. (1991). Проучване на могила 4 от източния могилен некропол в Сборяново .—АОР за 1990,46-47)

Михайлова Ж. (2005). Глинени огнища и олтари от тракийския град в Сборяново (предварителни наблюдения). МИФ 9, 229-240

Овчаров Д. (2006). Декоративни орнаменти или сакрални знаци в текето Демир баба при град Исперих.- Паметници, Реставрация , Музеи, декември 2006, 3-7

Фехер Г. (1935). Могилни находки от Мумджилар.- ИБАИ VIII, 106-114.

Чичикова М. (1975). Жертвеники еллинистический эпохи в Фракии.-Studia Thracia 1, Sofia 180-195.

Чичикова М. (1990). Новооткрит епиграфски паметник на култа на Фосфорос в Североизточна Тракия.- TAB, IV,82-92

BALKANSKA, A. (1998). *The Thracian Sanctuary at Demir Baba teke. "Sboryanovo" series, II.* Sofia

DIMITROVA-MILCEVA, A. (1990). Sacrificial altars of the Hellenistic Age in Kabyle. In: *Terra Antiqua Balcanica* , Sofia, 119-125.

DIMITROV, D.P., CICIKOVA, M. (1978). *The Thracian ciry of Seuthopolis,* Oxford, BAR, Suppl. Series 38

FOL, Al., CHICHIKOVA, M., IVANOV, T., TEOFILOV, T . (1986.) *The Thracian tomb near the Village of Sveshtari.* Svyat Publishers Village of Sveshtari. Svyat Publishers

DI FRAIA, T. (2012). A case of prehistoric incubatio: The cult site of Civitaluparalle, Chieti, Italy. In: Meaden, G. T. (2012) *Archaeology of Mother Earth Sites and Sanctuaries through the Ages: Rethinking symbols and images, art and artefacts from history and prehistory.* British Archaeological Reports. Oxford.

GERGOVA, D. (1988). Die Bedeuting der Schatzfunde in der Thrakischen Welt. In : *Der Thrakischen Silberschatz aus Rogozen, Bulgarien.* Katalog. Bonn-Mainz-Freiburg-Munchen-Hamburg, 1988, 51–58.

GERGOVA, D. (1988a). *The Thracian site at Kamen Rid locality near the village of Malak Porovets.-Excavations, Results and Problems. TAB III,*165-172

GERGOVA, D. (1995). The Treasure from Vulchitran and the Amber Route in the Balkans.-In: *The Early Hallstatt Period (1200-700 B.C.) in South Eastern Europe. Alba Julia 1994* . – Apulum, 1995, 69–79.

GERGOVA, D. (1996). The Rite of the Immortalization in Ancient Thrace. (=Гергова, Д., Обредът на обезсмъртяването в древна Тракия with English Summary, София, 1996, Агато.)

GERGOVA, D. (2004). *Sboryanovo—The Sacred Land of the Getae.* Bulgarian Bestseller. Sofia, 84 p.

GERGOVA, D. (2006). L'eschare dans le monde Thrace et Celte. – In: *Thracians and Celts. Proceedings of the International Colloquium from Bistrita, 18-20 May 2006.* Editura Mega,Clij-Napoca, 149–166

GERGOVA, D. (2007a). Sanctuaires et places de culte des gètes dans la Bulgarie du Nord-Est. In: *Iron Age Sanctuaries and Cult Places in the Thracian World. Proceedings of the International Colloquium Brasov, 19-21 October 2006.* Brasov, 63–78.

GERGOVA, D. (2007b). Les tombaux Thraces de la Bulgarie de Nord–Est. *L'archéologue [Paris], № 90, Dossier Les thraces*, 8–11.

LAZOV, G. (1996). *Decorated Clay Altars.* Pistiros I. 63-73

MAKIEWICZ,T. (1987). *Formy kultu bostw domowych na terenie Europy w starozytnosci.* Poznan.

Fr. MARTY, M. DEL CORSO (2002). L'habitat de hauteur du Castellan (Istres, B. du Rhone) à l'Age du Fer. *Etude des collections anciennes et recherches récentes. D A Med* 25, 129-169.

MEADEN, G. T. (1991) *The goddess of the stones: The language of the megaliths.* Souvenir Press, London.

STOYANOV, T. (1992). The Early Iron Age necropolis at the Singers' Field locality. In: *Sboryanovo series, vol. I, Sofia*

STOYANOV, T., MIHAYLOV, Z., NIKOV, Kr., NIKOLAEVA, M., STOYANOVA, D. (2006). *The Getic Capital in Sboryanovo.* Sofia.

Medieval Simulacra of the Mother Earth in the Christian Traditions

Orhideja Zorova

Institute for Archaeology and History of Art,
Faculty of Skopje,
R. Macedonia

Abstract. *Having in mind that throughout the Medieval Period the Euro-Asian territory—namely, the borders of the Christian cultural sphere—suffered numerous administrative and political changes (such as the forming of new states and the collapse of several old states), we cannot possibly account for and explain every different manifestation of the religious beliefs of medieval man. Nevertheless, the ancient origins of Christianity maintained some universal elements concerning the representation of one primordial/prehistoric concept, where the divinity was transformed into a personification.*

The Mother Earth, once venerated as a primal, if not only, deity in the Medieval Period became just a remainder of one ancient magnificent goddess. Henceforth, her existence remained on the margins of the official doctrine, or within the esoteric domains of what is known as mysticism in Christianity and in the superstitious behaviour of ordinary people. The goal of this paper is to offer a list of testimonies concerning the existence of the Mother Earth's effigy or image through different Christian cultural and territorial contexts of the Medieval Period.

Because this period is known by more than just archaeological evidence, we feel obliged to present its surviving clues in other media also, such as art, religious writings, and architectural and geomorphic forms in some way connected to the concept of Mother Earth. Thus in presenting as many examples as possible, the paper will nonetheless focus on those having a most provocative iconographic and religious manifestation. Emphasis will be given to examples of the material and pictorial sphere where the geo-anthropomorphic or the symbolical visualization of the Mother Earth is present.

The Medieval mundane view would be incomplete unless the two most important female characters of Christianity—Sophia and the Virgin—are mentioned. Beside their canonical identification and interpretation, the medieval sources offer yet another, apocryphal conception of these characters: Sophia as a "Mother of All Creation" (i.e. the material world) and the Virgin as a protector and benefactor of the deceased. In this way they become separate Christian hypostases of the functions once possessed by the Mother Earth.

We expect that through this paper some evidence concerning the transformation and demythologization of the deity will come to light, showing that the Mother Earth did not disappear through the Medieval Period. She simply accepted de-divinization, and continued her existence by way of simulacra.

For the full publication refer to pages 325 to 332 of the following:

http://www.ni.rs/byzantium/doc/zbornik8/PDF-VIII/23%20Orhideja%20%20Zorova.pdf

accessed 20 April 2012

Stones Called Baba as a Continuous Symbol of the Female Divine Principle: Folklore, Mythology and Archaeology

Katya Hrobat

University of Ljubljana, Faculty of Arts,
Department of ethnology and cultural anthropology,
Aškerčeva 2, SI – 1001 Ljubljana, Slovenia
katja.hrobat@guest.arnes.si or katja.hrobat@ff.uni-lj.si

Abstract*. This contribution analyses anthropomorphic megaliths, rock cliffs and the toponyms called Baba (which means Hag, an elderly wise-woman) from the point of view of folklore, mythology and, where applicable, archaeology. The image of Baba conceals a representation of an old woman, which is—in the case of Rodik (Slovenia)—manifested in natural stone as the image of a woman with exaggerated attributes. That we are dealing with such a representation of a woman can be proven with an instance of Baba at Grobnik (Croatia) which is cut into the wall at the entrance to the town. In the vicinity of these so-called monoliths, rock cliffs or toponyms (place names), specific rituals were performed. Baba stones were in one way or the other connected with water. According to folklore across the Slavic world, Baba seems to be an omnipresent phenomenon, connected to atmospheric phenomena like precipitation (rain, snow, and hail), rain clouds, sudden frost, drought, celestial objects, days and periods, spinning etc.*

Because of the large extent but also certain vagueness of the folkloric material on Baba, the contribution focuses on the spatial contexts of the Baba monoliths. What can be discerned is a simple repetition of the configuration of Baba connected with moisture, and cultic points linked to the notion of fire. The second starting point has its origins in an interesting custom from the Slovenian and Croatian coastal region—a threat made to children—that it is necessary to kiss the Baba and blow into her behind on the first visit to a certain town.

It is interesting to note that Babas are frequently positioned in the immediate vicinity of archaeological settlements, from prehistory onwards. Despite the fact that the tradition about Babas has a strong presence in the Slavic world, Babas—as anthropomorphic monoliths—are often found at those places that bear traces of even older traditions. Even the very stories about Babas can be linked with pre-Slavic traditions, among others Greek and Egyptian traditions also. One cannot overlook the impression that—due to their material stability—stone monoliths invite and attract the female divine principle, which in one way or the other is continuously preserved throughout all epochs up to the present day

Dust of the Ancestors: the Essence of Being and the Divinity of the Earth

Roberta Robin Dods

Community, Culture and Global Studies
Irving K. Barber School of Arts and Science,
University of British Columbia, Okanagan, Canada
robin.dods@ubc.ca and rrdods@ouc.bc.ca

***Abstract**. The place of humans in the 'natural' world has always caused comment and not just with the anthropologist or the human ecologist. As Von Maltzahn observes: "Humans do not have a world that is fixed for their kind; instead, there is a vital experience of space...The vital functions of a body subject are carried out within a vital space, which is not given as such but is shaped by that being's actions" (Von Maltzahn 1994:67). These actions are based in belief systems and what we, living in the Western Tradition, term the sentient world may be interpreted through sets of mythological and pragmatic constructs that give meaning to humans and those other than human as "... animate beings in motion against the backdrop of a terrain that... (is)... itself continually in process..." (Riddington 1982:473). This paper examines concepts of the sacred being-ness of the earth with emphasis on ideas of the divine that are centred in concepts of the feminine. Examples will be introduced from indigenous religions of the Americas to illustrate specific ideas about our link to what some term Mother Earth. Such concepts lead us to a deeper understanding of landscape analysis as it can be applied to archaeological cultures of deep time.*

References

DODS, R. Robin. 2003. Wondering The Wetland: archaeology through the lens of myth and metaphor in Northern Boreal Canada. *Journal of Wetland Archaeology* 3: 17-36. Oxbow Books.

DODS, R. Robin. 2004. Knowing Ways /Ways of Knowing: Reconciling Science and Tradition. *World Archaeology* 36:4 (Dec.2004). Pp. 547-557

RIDDINGTON, R. 1982. Technology, World View, and Adaptive Strategy in a Northern Hunting Society. *Canadian Review of Sociology and Anthropology*, Vol.19(4):469-481

VON MALTZAHN, K.E. 1994. *Nature As Landscape: dwelling and understanding*. Montreal: McGill-Queen's University Press

Earth and Sacredness in the Contemporary World: A Study Case

Dragos Gheorghiu

Department of Postgraduate Studies,
National University of Arts, Bucharest
gheorghiu_dragos@yahoo.com

***Abstract**: The ethnographic example presented in this paper shows the process of the transformation of a contemporary land-art work into a sacred symbol of the Earth, worshipped today by an Indian community near Bangalore. A geo-anthropomorph with obstetrical shape was painted by a contemporary artist and as a consequence perceived as a rhetorical trope of the goddess Shakti, as well as a pantheistic hierogamy of Earth and Sky, after being subjected to a ritual of consecration.*

***Key words:** Earth, hierogamy, land art, Mother Earth, religious symbolism.*

Introduction: Being and Earth

Parallel to the material archaeological record, there is an immaterial one, containing symbolic and metaphoric images, which is the imagery (Bachelard 1943; Burgos 1998)—and among the basic symbols of the traditional imagery are those referring to Earth.

Philosophers and anthropologists approached this subject during the last century (Bachelard 1948; Eliade 1949; Durand 1969) but it is only recently that one can observe an interest towards an "archaeology of Earth" (see for example Gansum 2004), discussing the Earth in connection with the Sky (Bender 2009), for example.

There is rich historical epigraphic and ethnographic heritage, which demonstrate that "Earth", or "Mother Earth", is almost a universal symbol (see extended bibliography in Eliade 1949), and that the cognition and identity of people living in traditional societies depended on their connection with the Earth, "the ever-sure foundations of all" (*Theogony*...1914):

> "The earth is my mother, and I the son of the earth" (*Hymns of the Atharva Veda*,
>
> XII, v.12, cited in Eliade 1977: 38).

> "I will sing of well-founded Earth, mother of all, eldest of all beings" (*The Homeric*
>
> *Hymns* 1914 XXX, cited in Eliade 1977: 55).

In this perspective, the problem I discuss in the present paper is whether contemporary imageries still use symbols of the Earth, and whether in the present-day world there are still places of cult and sanctuaries of Earth or Mother Earth.

Earth – an archaic symbol

According to Mircea Eliade (1949), the archaic symbols of the Greek cults of Earth (or Gaia) seem to have been overlapped by more recent ones, specific for theocratic states and patriarchy, which worshipped Mother Goddess images (Morris 1998; Morris 2006). Common to both periods is the same basic symbolism, i.e. the feminine divine.

This perception of female sacredness had a long tradition in the Western thinking of Antiquity (Noyce 2007), as well as in antique and contemporary Eastern thinking (Bhattacharyya 1999). In Western Europe, the concept of the divine feminine re-emerged in 19th-century high culture (Bachofen (1861/1967), and then at the beginning of the 20th-century with the Russian mystics, to cite only V. Soloviev and S. N. Trubetskoi (Groberg 1998).

Although there was an interest in high culture to approach the subject, its study was neglected in folk cultures. Practically all around the world in folk cultures, like in the ancient societies, the couple, and not a single deity, played a central symbolic role in geneses and the survival of the universe (Eliade 1949: 215); as a result, the symbol of a female Earth (i.e. a Mother Earth) was worshipped as part of a sacred couple:

> "The earth [...] mates with Indra the bull" (*Hymns of the Atharva Veda* 1891: XII, v.6,
>
> cited in Eliade 1977: 38);

> The earth "lay with Sky and gave birth to Ocean"
>
> (Hesiod 1953, cited in Eliade 1977: 117)

or

> The Earth is "Mother of the gods, wife of starry Heaven"
>
> (*The Homeric Hymns* XXX).

Geo-anthropomorphs

One of the universal cultural processes is anthropomorphism (Fisher 1991; Mithen 1998), and is the reason why in the imagery of many traditions Earth appears in relation with anthropomorphic shapes (for the Mediterranean area see for instance Pausanias 1918 or The Homeric Hymns 1914), consequently there was an analogy between parts of the female body anatomy and shapes that resulted from Nature's processes.

The conversion, in different contemporary communities from the Mediterranean area of various geological forms, derived from the leakage of pluvial waters through soft rocks or stones resembling the female genitals (Fig. 1), into communal monuments (Figs. 2 & 3). This is comparable to an act of consecration, and implies that there are still similarities between modern imagery and that of the past. Such a kind of geo-anthropomorphic analogy was current in Antiquity, because "numerous populations located the gestation of children in grottoes, in rock slits, as well as in springs" (Durand 1969: 285) (Fig. 4).

I shall illustrate this similarity between ancient and contemporary imagery, with an ethnographic example from India, which supports the idea that Earth, through geo-anthropomorphs, continues to be a generator of sacred symbols.

The symbol of the primordial couple as a geo-anthropomorph

The example mentioned relates to popular Hinduism, and represents a land-art made in 1999 by Pravin Cherkoori near Bangalore, on the Deccan Plateau, in the south-eastern part of the Karnataka state in India.

Pravin Cherkoori (born 1932), a Hindu visual artist based in Vienna, inspired his abstract art from popular Hinduism, a recurrent theme being the representation of the divine couple of Shiva-Shakti, imagined, like in folk art, as infinite avatars, from three-dimensional forms like Lingam and Yoni, to colour codes (see Gheorghiu 2005: 72ff.). In many of his artworks, as in folk art, the female goddess is symbolized by red dye, and the male god is symbolized by blue dye. Put together the two colours represent the materiality of the universe (i.e. Earth and Sky).

In popular Hinduism (Fuller 1992) Shiva's consort, Parvati—in vernacular terms matta or ammai (mother), or "Shakti, expressing female power, energy and

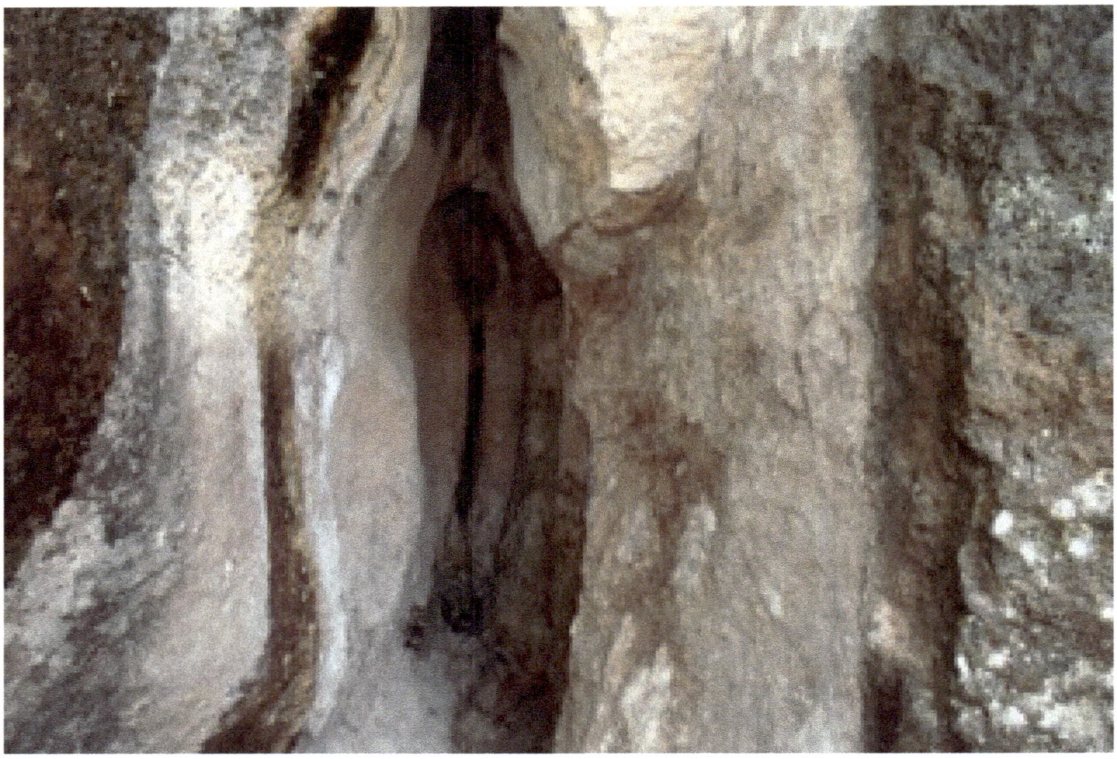

Fig. 1. Coralline limestone geomorph near at the entrance of Għar Dalam cave, Malta

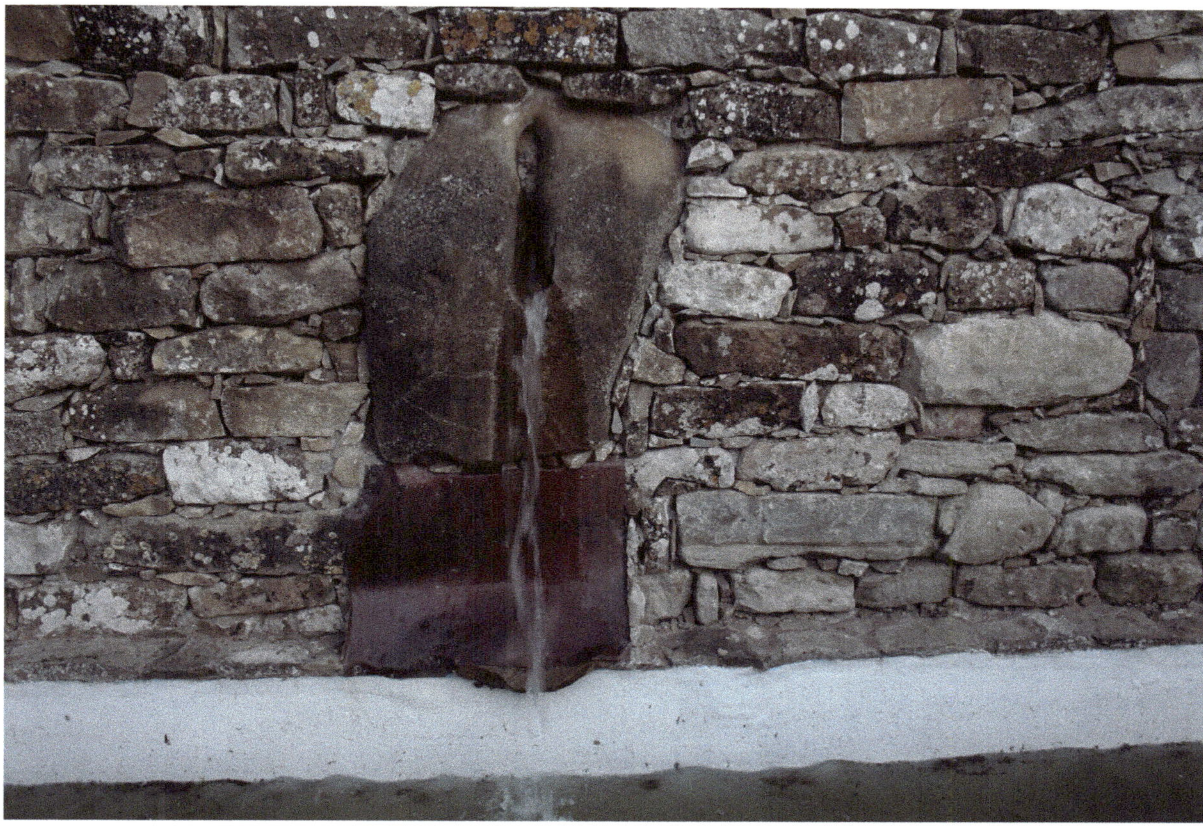

Fig. 2. Public fountain made from a fragment of a geo-anthropomorph, Mandas, Sardinia.

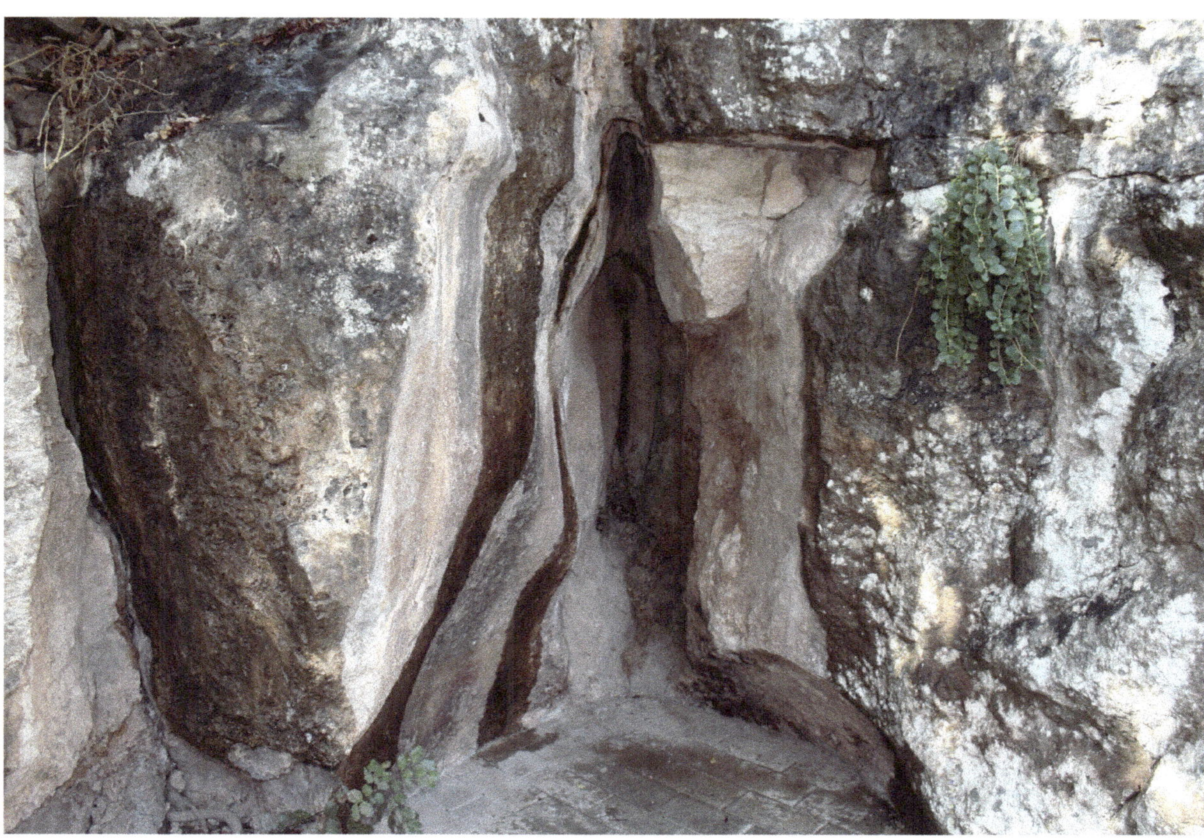

Fig. 3. Modern pavement to approach a geo-anthropomorph, Għar Dalam cave, Malta. One can see the intervention at the base of the rock to allow the viewer to come near to the inner details.

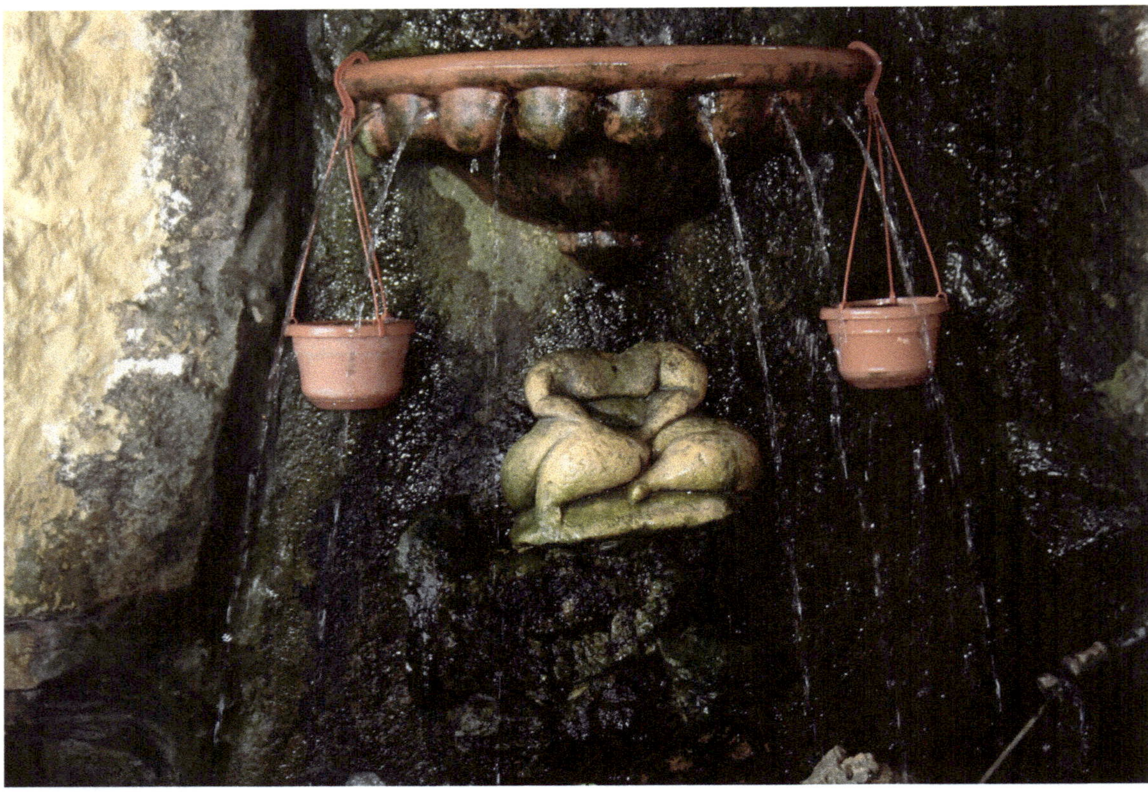

Fig. 4. A replica of a female figurine from Ħaġar Qim positioned on a fountain. Malta.

independence" (Morris 2006: 124)—is the mother goddess who returns to people the transmuted offerings in the form of shakti, i.e. a sacred power or blessing (Preston 1980), similar action to that performed by the Mother Goddess in the Vedas ("O mother earth [...] place me into happiness and prosperity", Atharva Veda XII, 63), or in the Homeric Hymns (Evelyn-White 1914) ("Happy is the man whom you delight to honour! He has all things abundantly, etc.")

To visualize in a modern form the divine couple Shiva–Shakti as a work of art at a grand scale in Bangalore, Pravin chose a high rock with a slit centrally positioned (Fig. 5), and which had a boulder on its top. The slit that resulted from water erosion created a suggestive image of the female genitals, especially after Pravin coloured its contour in red (Fig. 6). To emphasize the rituality of his gestures, the artist realized a series of rows of white dots resulted from a standardised movement of the hand on the stone surface (Fig. 7). The resulting work had no connection to popular Hinduism art representations of the divine couple, being an original work of Modern art (Fig. 8).

If in European and Indian high cultures this land-art was received as a contemporary art production, it functioned for the local population as a religious symbol, the work receiving sacred connotations as being a darshana (i.e. the vision of a deity, Morris 2006: 132) and then a murthi (i.e. an embodiment of the divinity in matter, after the divine was evoked in it, Nagendra 1997: 739), because in popular Hinduism the objects of worship are the deities hidden inside the image produced (Fuller 1992: 60).

After the completion of the land-art, the place became sacred, the local people sending prayers to the geo-anthropomorph which represented for them not only the Shiva-Shakti symbol but also a Vedic pantheist epiphany of the matta / ammai Earth and father Sky (Fig. 9).

The Vedas underlined this relationship between Earth and human religious activities:

> "The earth upon which they (the priests) enclose the altar (vedi) upon which they (...) unfold the sacrifice (...) that earth shall prosper us, herself prospering" (Atharva Veda XII, 13).

It is well known that popular Hinduism is characterized by a tendency to pantheism (Srivinas 1962) and to syncretism, eliminating a neat difference between great gods and village local deities (Fuller 1992: 55). Because in Tantric Hindu imagery the female genitals or yoni (see Morris 2006: 123 for this representation of Goddess) represent a sacred image or sign, the artistic intervention on the Bangalore rock which emphasized that semiotic acquired sacredness and consequently was worshipped (puja, Fuller 1992).

Conclusion: the survival of the traditional symbols

The Bangalore example showed how geo-anthropomorphs (i.e. Earth) could evoke symbols from the natives' imagery if they are revealed to the viewers in a ritual way. Here, after being revealed to the community following the ritual of colouring (which was received by the community as a

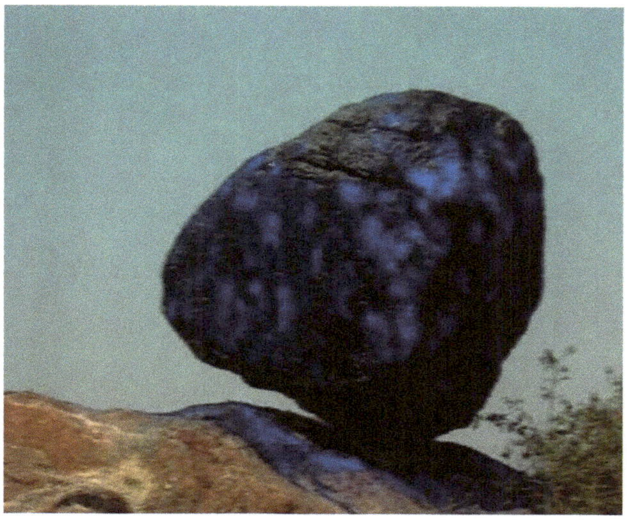

Fig. 5. Boulder positioned on a rock with a central slit, Bangalore, India (still image from film).

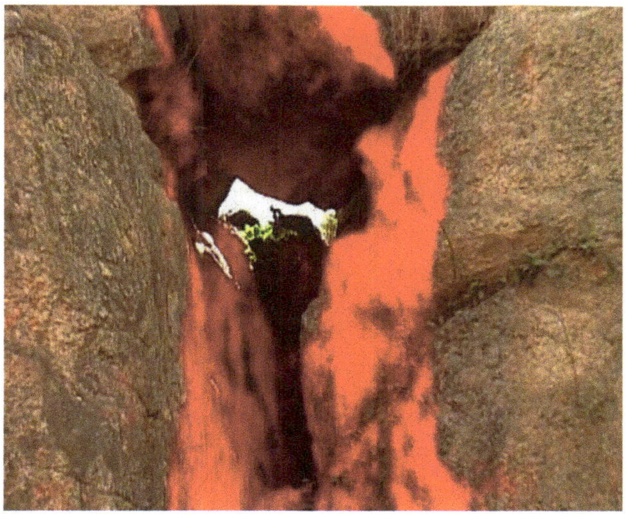

Fig. 6. The slit of the rock painted in red (still image from film).

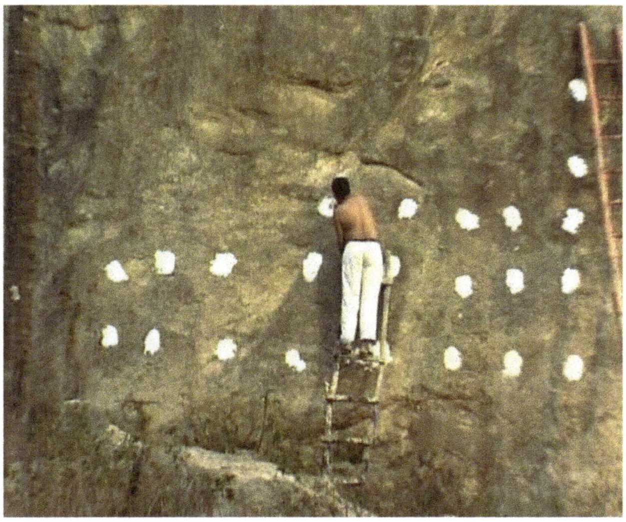

Fig. 7. Pravin Cherkoori painting the white dots on the rock surface (still image from film).

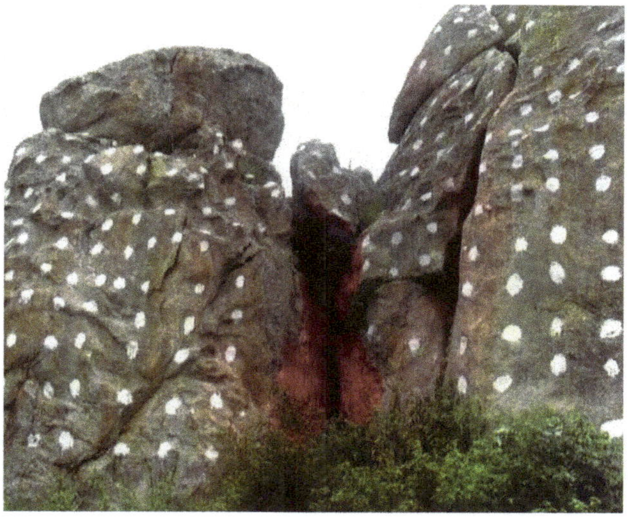

Fig. 8. The Bangalore rock after completion of the land-art (still image from film).

ritual of consecration or prāna-pratişţhā when the image of a deity is animated through an elaborate ceremony, Lipner 1998: 280), the slit in the rock was perceived as the synecdoche of the female body, which in turn was perceived as an index of the hierogamy that regenerates and fertilizes the world, fertility being one of Earth's attributes (see Hymns of the Atharva Veda, v. 4-5). From that moment the geo-anthropomorph was received in the imagery of local people as a darshana, or a natural sanctuary (see Bradley 2000).

Therefore the sacredness of a place depends on the capacity to in-materialize (or, from an anthropomorphic perspective, to embody) and to ex-materialize (or to reveal) the sacred, a quality which is specific for stone (Eliade 1949 gives an extended bibliography on the hierophany character of this material) and consequently to Earth.

This character of the Earth to embody sacredness is underlined in the description of the deities' embodiments in different earthly materials: "the gods entered their bodies, of all kinds of wood, all kinds of minerals, all kinds of clay … in which they took form" (Morentz 1973: 154, cited in Meskell 2006: 55). In the Bangalore case after the consecration, the rock embodied shakti the female power and energy.

I conclude this paper with the assertion that the processes of anthropomorphization, as well as that of epiphany and hierogamy, which were specific for traditional societies, were not totally withdrawn by Modernity, and still could exist unrevealed within contemporary imageries.

Acknowledgements: The author thanks M. Pravin Cherkoori for the kind help with the film of the Bangalore land-art and for the data concerning the making and reception of the monument. The photographs of Figures 1- 4 are by the author.

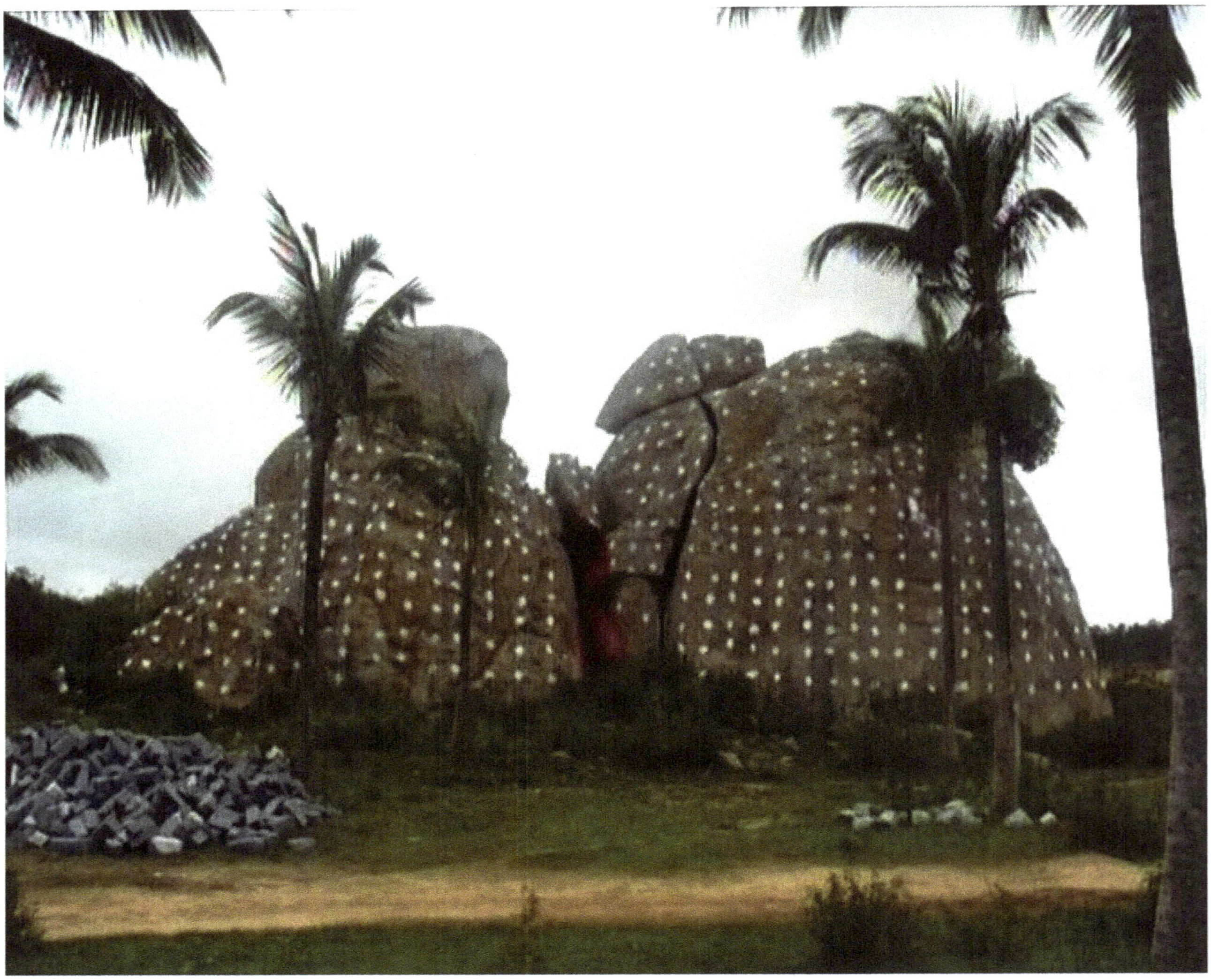

Fig. 9. Full view of the painted Bangalore rock after completion of the land-art (still image from film).

References

BACHELARD, G. (1943) *L'air et les songes. Essai sur l'imagination du mouvement*. Paris, Librarie José Corti.

BACHELARD, G. (1948) *La terre et les réveries de la volonté*. Paris, Librarie José Corti.

BACHOFEN, J.J. (1967) *Myth, religion and Mother Right*. Princeton NJ, Princeton University Press (Originally published in 1861).

BHATTACHARYYA, N.N. (1999) *The Indian Mother Goddess*. New Delhi, Manohar.

BENDER, H.E. (2009) Caring for creation – A hierophany at Strawberry Island, pp. 137- 156. In Nash, G. and Gheorghiu, D. (eds.), *The Archaeology of people and territoriality*. Budapest, Archaeolingua.

BRADLEY, R. (2000) *An Archaeology of natural places*. London and New York. Routledge.

BURGOS, J. (1998) *Imaginaire et Création*. Paris, J-P. Huguet.

DURAND, G. (1969) *Les structures anthropologiques de l'imaginaire*. Paris, Bruxelles, Monréal, Bordas.

ELIADE, M. (1949) *Traité d'histoire des réligions*. Paris, Payot.

ELIADE, M. (1977) *From primitives to Zen. A thematic sourcebook of the history of religions*, San Franciso, Harper and Row.

FISHER, J., A. (1991) Disambiguating anthropomorhism: An interdisciplinary review, pp. 49-85. In BATESON, P.P.G., and KLOPFER, P.H. (eds.), *Perspectives in ethology*, Volume 9, Plenum Publishing Corporation.

FULLER, C.J. (1992) *The Camphor flame*. Princeton NJ, Princeton University Press.

GANSUM, T. (2004) The Archaeology of Earth. *Current Swedish Archaeology*, Vol. 12, pp. 7-21.

GHEORGHIU, D. (2005) *An Archaeology of art. Pravin Cherkoori- A study*. Vienna, Bondi Verlag.

GROBERG, K. (1998) Vladimir Sergeevich Solov'ev: a bibliography. *Modern Greek Studies Yearbook* 14-15, pp. 299-398.

HESIOD, *Theogony* (1955) pp. 314-316. In Zaehner (ed.), *Zurvan: A Zoroastrain dilemma*, Oxford.

The Homeric Hymns, 1914, p. 456. In Evelyn-White, H.G. (ed.), New York, Loeb Classical Library.

Hymns of the Atharva Veda (1891) translation by Maurice Bloomfield, Oxford.

LIPNER, J. (1998) *Hindus. Their religious beliefs and practices*. London and New York, Routledge.

MESKELL, L. (2006) Objects in the mirror appear closer than they are, pp. 51-71. In Miller, D. (ed.), *Materiality*, Durham and London, Duke University Press.

MITHEN, S. (1998) *The Prehistory of the Mind: A Search for the Origins of Art, Religion and Science*, Phoenix.

MORENTZ, S. (1973) *Egyptian religion*. Ithaca N.Y., Cornell University Press.

MORRIS, B. (1998) Matriliny and Mother Goddess religion. *Journal of contemporary religion* 13 (1), pp. 91-102.

MORRIS, B. (2006) *Religion and anthropology. A critical introduction*. Cambridge, Cambridge University Press.

NAGENDRA, K.S. (1997) *Encyclopaedia of Hinduism, Volume 7*. Anmol Publications Pvt. Ltd., Centre for International Religious Studies.

NOYCE, J. (2007) *The Wisdom tradition, Visions and prophecies of the Goddess in the Sapiential tradition*. Melbourne, Noyce Publishing.

PAUSANIAS (1918) *Description of Greece*. Translated by Jones, W. H. S. and Omerod, H. A. Loeb Classical Library Volumes. Cambridge, MA, Harvard University Press; London, William Heinemann Ltd.

PRESTON, J. J. (1980) *Cult of the Goddess*. New Delhi, Vikas.

SRIVINAS, M.N. (1962) *Caste in Modern India and other essays*. London, J.K. Publishers.

The Theogony of Hesiod (1914) translated by Hugh G. Evelyn-White.

Index

www.ingramcontent.com/pod-product-compliance
Lightning Source LLC
LaVergne TN
LVHW071219210926
848482LV00015B/849
* 9 7 8 1 4 0 7 3 0 9 8 1 1 *